T0228189

Strategic Management of Research Organizations

Strategic Management of Research Organizations

William A. Barletta

Department of Physics
Massachusetts Institute of Technology

Faculty of Economics
Ljubljana University

CRC Press is an imprint of the
Taylor & Francis Group, an **informa** business

CRC Press
Taylor & Francis Group
6000 Broken Sound Parkway NW, Suite 300
Boca Raton, FL 33487-2742

© 2020 by William A. Barletta

CRC Press is an imprint of Taylor & Francis Group, an Informa business

No claim to original U.S. Government works

Printed on acid-free paper

International Standard Book Number-13: 978-0-367-25585-5 (Hardback)

Visit the Taylor & Francis Web site at
http://www.taylorandfrancis.com

and the CRC Press Web site at
http://www.crcpress.com

Contents

Foreword   ix
Acknowledgments   xiii
Author   xv

1 Perspectives and Networks   **1**
The Enterprise Planning Perspective   1
Operational Networks   5

2 Measures of the Research Manager   **9**
ELM Space   9
Leading Creative Employees   12
Managing Organizational Change   16

3 The Research Environment   **19**
What is Research?   19
Trends in Research Funding   21
The R&D Life Cycle   22
Your World as an R&D Manager   23

4 Strategy, Forecasting, and Technological Risk   **27**
Characteristics of a Business Strategy   27
Enterprise Risk Management   29
Technological Forecasting   31

5 Introduction to Strategic Planning   **37**
Preparing to Craft the Content of the Plan   37
Phase 1—Gathering and Assessing Information   38
Corporate Strategies and Positioning   42
Crafting the Plan Document   44
Resource Planning for Implementation   46
Information Systems   48

6 Financial Management   **49**
Basic Accounting Definitions   50
Tools for Financial Management   52

Costs as a Decision-Making Tool   55
A Manager's Responsibility   56

7 The Business Plan   **59**
Business Opportunity   61
You Have a Great Idea, but How Do You Plan to Make Money?   63
Writing the Business Plan   64
Why Do Plans Fail?   67

8 Management Communication Skills   **69**
Technical Writing   69
General Comments about Style   70
Major Formal Writing Projects—The CDR   72
Management of Meetings   76
Negotiations (in Brief)   77
Investigation for Managers   80
The Manager as Judge   80

9 Marketing Scientific Organizations   **83**
Marketing for Sales: Identifying a Product Line   83
Marketing for Sales: Positioning Products   86
Strategic Marketing   88

10 Research Ethics   **91**
Ethical Issues of Scientific Research   93
Conflicts of Interest   96
Institutional Ethics   96

11 Workforce Management   **99**
Building Your Workforce Strategically   99
Hiring the Best   101
Succession Planning   103
Salary Management   106

12 Managing Operating Risks   **111**
Controlling Risks during Execution   111

13 Structures and Governance   **117**
Organizational Structures   118
Institutional Governance   122

14 Technology-Transfer Case Study   **125**
     KYMA Case Study   126
     Keys for Lasting Success of Technology Transfer   128
     Strategic Partners and Allies   129

15 Recommended Resources   **131**

Index   135

Foreword

The material for this text has grown from a course that I developed and taught for several U.S. universities and for the University of Ljubljana. I envision the target audience to be middle managers who aspire to top management positions at large research laboratories and to technology entrepreneurs who envision growing their small companies into a substantial corporation. Scientists and engineers who are promoted to management positions in major research organizations are most typically selected because of their technical acumen rather than their executive, management, or supervisory skills. Typically, the most common training provided by their organizations concentrates on "leadership" and selected transactional skills connected with the organization's Human Resources (HR) department. Any broader perspectives are frequently part of training in project management. Such courses can be valuable, yet developing the manager's skills in strategic thinking is quite frequently absent—especially from an enterprise-wide perspective. While such tactical training may suffice for first-line managers and supervisors, it generally falls far short of what is needed to produce talented senior and upper-middle-level executives who need to deliver maximum value to their organization as a whole. As the size and cost of major research infrastructures grow ever larger, the lack of holistic, executive training is becoming ever more serious. The author has developed the course material in this book to fill this gap and to serve as an introductory overview for an MBA program focused on the strategic management of research enterprises.

When I began preparing for my first course in this area, "Managing Science in Research Laboratories," sponsored by Texas A&M University in 2007, I thought that the process would be rather simple. I would just codify what I had been doing as a senior executive and line manager in an American government national laboratory for the previous 20 years. However, as soon as I started looking through the literature for additional examples outside my direct experience, I found that the range of ideas and the complexity of the many diverse approaches required both additional research and a careful systematization of thinking about the question, "In what ways are managing scientific and engineering research enterprises different from managing a factory or a service enterprise?"

In simply formulating this root question, I repeatedly found the word "enterprise" coming to the fore. I was already familiar with the slogan-like use of "enterprise" to describe approaches to cyber-security and other operational mandates in the private commercial sector. That usage had a distinct purpose, namely, to emphasize that certain practices and approaches should be employed holistically across the entire organization. That realization served as the seed for the underlying organizing principle for my course. I emphasize that the management approach—especially for senior executives—should be enterprise-wide wherever practical, and it should be derivable from the top-level strategic view of the enterprise and carried through to day-to-day operations.

In crafting the lessons described in this book, I have adopted three distinct points of view:

1. The strategic or executive viewpoint. This point of view is top-down "peeling the onion." It assumes that the top management has a clear view of the organizational positioning of their enterprise. For many research enterprises, a natural organizational positioning is that of the technology leader.
2. The managerial or operational point of view. This viewpoint examines enterprise operations proceeding from the focus on the customer to the focus on high-quality performance at all levels of the enterprise.
3. The technology-driven viewpoint. This point of view is especially important for managers of major research infrastructures. It is informed by the end-user of the infrastructure or its technology. It seeks to establish quantitative (or at least semi-quantitative) criteria or figures of merit. These figures of merit can also serve as surrogates for the financial bottom line of a commercial business.

The examples in the text and online supplementary materials that are related to these points of view are drawn from my background as research executive, as an operations manager, as a research physicist, and as a technologist.

The reader may wonder why research enterprises and their managers and executives need systematic training in strategic management. I see four principal reasons:

1. Laboratories and universities have an objective need because (1) modern team science is complex; (2) we don't train scientists and engineers on how to manage; and (3) being competitive means delivering more "bang for the buck."

2. Mistakes are increasingly costly to the organization: (1) there is heavy competition for funds; (2) law suits are frequent, costly, and time consuming; (3) funding agencies are increasingly intolerant of risk; (4) for government laboratories and universities, institutional embarrassment is a significant risk.

3. Good managers should also improve science/technology (1) to maintain their scientific/engineering credibility, (2) to build better research institutions, and (3) to maintain their professional pride.

4. Our stakeholders have high expectations of us as managers. For all organizations, this means assuring worker and public safety, respecting the environment, and maintaining public trust. For commercial enterprises, expectations include financial success and gaining market share. For public institutions and universities, expectations also include scientific and technological policy criteria such as *academic excellence, excellence of research staff*, and *pervasive societal benefits*.

Over the years, I have found that the lessons learned or the pieces of advice I remembered best are those that I've been able to distill into a short sentence or aphorism. Throughout this text, I have inserted "Theorems" and "Corollaries" where they can effectively encapsulate a succinct lesson learned from that chapter. The process of deriving "theorems" begins with looking for trends, remembering simple examples, extracting the aphorism, and then drawing parallels with other types of enterprises. As a senior-executive colleague cautioned me many years ago in the form a "theorem" of his own, "If you can't write it on a tee shirt, you won't remember it." I'd add a corollary to that theorem, "And you won't convince others of the advice." The best such distillations are those that the listeners (or readers) derive and can apply personally to themselves. However, "theorems" are no replacement for developing one's instincts about people. If management were only about knowledge and algorithms, we'd let computers do it.

I have included figures sparingly in text, reserving them for visual summaries of major concepts or features that I consider to be especially novel in distinguishing the roles of the thought leader, the operations manager, and the executive. In practice, most research organizations require persons in the management hierarchy to embody some aspects of each of these roles. Therefore, the reader should ask him- or herself how the lessons apply to themselves personally as well as how they apply to their organizations.

Acknowledgments

Each time that I have presented my course, I selected a teaching partner who has been a colleague with extensive management experience. Their respective approaches to institutional governance, project management, personnel management, and best business practices have influenced much of this work.

One of these teaching partners stands out among all the rest. It is to Dr. Barbara Maria Thibadeau [Captain, United States Navy (retired)]—a superb project manager, an advocate of the highest standards of ethical practice, and an exemplar of natural command presence—that I owe the greatest thanks. Barbara worked as my financial officer for four years in Berkeley before moving to Oak Ridge National Laboratory where she completed her PhD degree in Business Management and served as a project manager. Barbara and I had intended to collaborate in the writing of this text. Alas, her untimely passing intervened. This book is dedicated to her memory.

I would also like to acknowledge the collaboration of Kem Robinson, David McGraw, Ajda Lah, and Luisella Lari in shaping course sessions upon which this text is based. I am grateful to my students who have encouraged me to write this text and to my wife, Beverly, who has supported me in my work for decades.

Author

William A. Barletta is Adjunct Professor of Physics at MIT and Director Emeritus of the Accelerator Division at LBNL. Internationally, he is Visiting Professor in the Faculty of Economics of the University of Ljubljana and Coordinating Editor-in-Chief of *Nuclear Instruments and Methods-A*, a member of the Scientific Council of the Centro Fermi and Museum in Rome, a member of the Advisory Board of the John Adams Institute in the UK, and senior advisor to the President of Sincrotrone Trieste.

He holds a PhD (Physics) from the University of Chicago and is a Fellow of the American Physical Society and a Foreign Member of the Academy of Sciences of the Bologna Institute in Italy.

Perspectives and Networks

1

THE ENTERPRISE PLANNING PERSPECTIVE

Vision statements and mission statements have become virtually ubiquitous in the corporate world. That practice has spread beyond commercial firms to research organizations from as small as single technology start-ups to large national and international government-funded laboratories. Nonetheless, beyond their value as feel good slogans, these obbligato texts frequently offer little actionable guidance to top and senior management. Moreover, for the rank and file worker, these statements—if they are even known—convey little meaning or sense of purpose to most employees. Why is that?

From the top-down perspective, these *foundational statements* are frequently developed *post-hoc* in response to pressures from shareholders, boards of directors, advisory committees, or government agencies. Although the post-hoc introspection may have some value to the management team, once the relevant pressure group has been satisfied, these statements recede into the background and may be easily forgotten. Yet they should reveal the starting point for managing strategically, as they are meant to describe "who we are" as an organization. This book starts with that perspective. Strategic management should be enterprise-wide with guiding principles and individual managers aligned with and driven by the mission and goals of the organization as a whole. That perspective embodies a superstructure of three tiers:

1. Who We Are (Mission, Vision, and Guiding Principles)
2. Our Goals, Strategy, and Plans
3. Our Understanding of the Market (Business) Environment.

For the research organizations, the market environment includes the research environment, broadly defined funding opportunities, and the potential for spinoffs (or commercialization).

The enterprise-wide approach merges smoothly into holistic management that integrates critical strategic activities and elements of *business operations* into a logical whole. Within the holistic perspective, the activities that embody the execution of the corporate strategy into management of operations (tactics) and that provide assurance and accountability to stakeholders both flow top-down in a logical way from the enterprise-wide superstructure. The *enterprise planning network* depicts the flow of logical connections from the space of executive authority to the space of operations management as shown schematically in Figure 1.1.

The execution (operations of the organization) can be considered to be ethical when "doing what is right" is also doing what is best for the organization in the long run.

In the Executive Space of Figure 1.1, the "business environment" encompasses the general research environment, the market structure, competitive and complementary organizations and forces, and the network of stakeholders relevant to the enterprise. The "policy environment" is especially important to those enterprises that receive more than a *de minimis* portion of their income (or support) from governmental or philanthropic sources. The policies, decisions, and subsequent actions of those sources strongly influence the financial

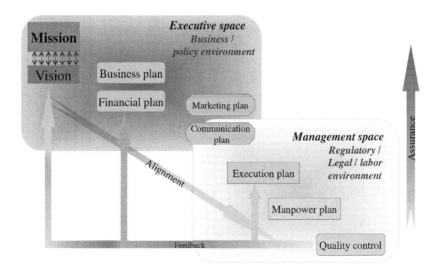

FIGURE 1.1   The flow of planning from strategy to tactics and from executive leadership to operations management.

viability of the enterprise in proportion to the fraction of its income from those sources. Naturally, the policies of disparate sources are frequently not aligned. Therefore, senior management must understand and navigate misalignments and conflicts. The top-level embodiment of that navigational process is the near- to intermediate-term business plan and its accompanying financial plan.

Finally, top and senior management must communicate its plans internally throughout the enterprise and also externally to its most important stakeholders and customers. Those later actions imply that the comprehensive scheme of Figure 1.1 omits a second network critical to the success of the enterprise. The *product-centric network* focuses the enterprise identity (the Why) and its resources (the How) plus the inputs of the customers and the investors onto the products of the enterprise (the What). For a research enterprise, the research output and its outward expression are an important, and often the primary, product line. The top and senior management must never lose sight of the products of its organization. Those products are the sum of the deliverables of the enterprise.

Topics for class discussion or reader introspection: (1) What other relevant networks are there? (2) Where and how does research fit in? (3) How does the enterprise decide what research to perform?

In the *Management Space*, the combined efforts of the enterprise managers must put the plans formulated and promulgated by the executives into action by the entire set of employees. To assure a cost-effective deployment of corporate resources, the managers of each business unit within the enterprise need an actionable plan with a near-term (3-year) time horizon. Such a plan should derive from an audit of available employee skills and result in a staffing plan for managers to assign (or hire) appropriate employees to take on specific responsibilities and to assure that their performance fulfills the objectives of the enterprise. These human resource plans need to recognize the availability of special skills in the labor market. They must be consistent with relevant labor law and representative contracts for unionized workers. To be legal and ethical, all operations must comply with regulations promulgated by all relevant levels of government. They should respect the environment and the health and safety of the public.

No employee should lose sight that the quality of the organization's products determines the long-term viability of the enterprise. In the end, assuring the quality of product as well the legal and cost-effective execution of the high-level plans of the enterprise is accomplished via multiple feedback loops of information to the appropriate senior executives. The organization can operate effectively to advance the enterprise strategy crafted by top and senior management when day-to-day operations are structured in a way that flows from and is consistent with the mission, vision, and guiding values of the enterprise.

The overall assessment of enterprise performance for which the *chief executive officer* is ultimately responsible is a bottom-up evaluation of performance assurance that holds each employee accountable for actions within his/her sphere of influence. Assurance reports are often the product of internal and/or external auditors in which overall performance is judged against the degree to which the enterprise is functioning cost-effectively to advance its mission in compliance with policy, regulations, and relevant law.

To work effectively, managers must feel themselves accountable (and being held accountable) for activities within their respective range of control. Performance-oriented employees can become good managers when they accept personal accountability for their actions, successes, and failures. It is often said that managers fail in direct proportion to their willingness to accept socially acceptable excuses for failure or to blame the levels of management above them with the excuse that "they made me do it." Likewise mangers must be willing to set and enforce standards of behavior and performance for their subordinates and for themselves. One cannot expect satisfactory assurance audits unless an environment of accountability pervades the enterprise.

If mission and vision statements are the ultimate yardsticks by which the governance body of the enterprise (for example, a Board of Directors) will judge the performance of top management, those statements must be more than anodyne. For major research infrastructures, the mission should offer a concise, compelling statement of why the infrastructure exists. For a technology start-up, the mission statement might contain a compelling description of providing the flagship product of the enterprise to the customer base.

Unfortunately many organizations craft vision statements that are little more than lofty aspirations about the future; i.e., their statements are not actionable. Some vision statements are mere restatements of the mission statement in more esoteric language. For example, one prominent national laboratory proclaims, "Our vision is to solve the mysteries of matter, energy, space, and time for the benefit of all." A taxpayer may wonder, "Since everything is either matter, energy, space, or time, what in the world is excluded from their vision? For eternity? And to what end?" Such a vision for a publicly funded institution is effectively asking for a blank check to do whatever the Laboratory Director General wants, as it does not provide stakeholders any quantitative— or even qualitative—criteria on which to make judgments and decisions. Vague language does not provide sufficient clarity for serious strategic planning; a collective of secular philosophers or ascetic monks might have the same vision.

To be actionable, a vision statement should state clearly and concisely where the enterprise is going, how it will get there, and what the time horizon for getting there is. For major research infrastructures, a 20-year planning horizon is a reasonable temporal horizon. For a small start-up, a 10-year vision would be appropriate.

Topics for discussion or introspection: (1) What is the mission of your laboratory, company, or university? What is the vision that is broadcast to external stakeholders? Are these statements actionable or are they merely blanket cover for whatever the CEO or lab director wants to do this year?

(2) What are common ways in which leaders mismanage people in research organizations? What kind of damage is done by such mismanagement? Are standards of performance clearly articulated in your organization? Are excuses commonly made for failures in your laboratory, company, or university?

OPERATIONAL NETWORKS

To analyze both internal and external organizational dynamics, a useful starting point is to describe the most important networks that drive the behavior of the enterprise either directly or indirectly. A few definitions and general concepts are in order at the outset. (1) A *network* is a collection of nodes. (2) A network has connections (links) between nodes. The links may be directional. They may be asymmetric; in order words, $a \geq b$ does not imply $b \geq a$. If the network contains nodes with a single link, it is said to be open. (3) The network has a set of rules that govern the links. In physical systems, these rules could be conservation laws.

a. Networks can describe social relationships such as the management organization, its reporting charts, proximity relationships in social networks, or "communities" and interest groups in which the nodes are actors.

b. Networks can describe physical connections. Examples include cryogenic and electrical circuits and computer networks. In these networks, nodes can be thought of as "solder joints." In electrical networks, the symmetric links are the linear circuit elements (resistors, capacitors, and inductors). Diodes would be asymmetric links. The network rules are Ohm's Laws.

  c. Networks can describe logical connections. In project management, logic diagrams describe which activities must precede others. Computer flow charts show the logical flow of processing information; in these networks, the nodes are milestones or decision points.

  d. Networks can be also tools of analysis, that is, highly idealized models of physical systems in which the nodes are topological.

The *opportunity environment* for managers and entrepreneurs is their local (social) network. This network is the map of the political landscape of the enterprise managers. The nodes of this network are the set of all potential stakeholders of the organization. In Figure 1.2, the dark solid lines are the *direct*, bi-directional communications and action links. The general direction of communication flow is (or should be) from bottom upward and from left to right. When employees complain in suggestion boxes and surveys that communications need improvement, they are generally lamenting the fact that the dark black lines are becoming ever more outwardly directed from the center. Unfortunately a common response of senior and top management is to double down on their deficient behavior, for example, by introducing a slick organizational newsletter or merely renaming the old one.

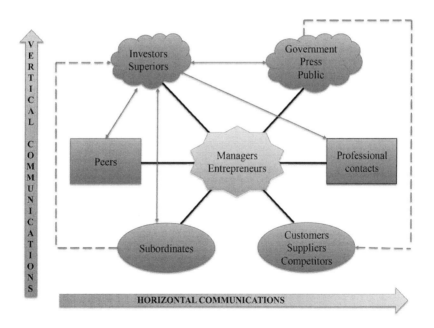

FIGURE 1.2 The local (political) network of enterprise mangers.

Management makes a substantial error when it ignores the secondary, indirect links in the network illustrated by the thin bi-directional arrows and dotted lines. These links can "bite" managers in multiple, damaging ways. For public sector organizations in which the consequences of public embarrassment can be severe and unpredictable, one should also include links from the bottom tier elements to the government and the press.

The higher one rises in the hierarchy of corporate management, the more critical that understanding the political network and its potential influences becomes. Although scientists and technical professionals often hear their colleagues proclaim that they hate politics, in failing to understand how to work successfully in their political network, these colleagues are undermining their own chances of success.

Savvy managers know that the network can provide a palpable source of power to those who may have no formal authority at all, even to outsiders. In this context, power means the ability to corral and mobilize resources (human, financial, and infrastructure) to get things done. Ideas alone, no matter how brilliant, change nothing.

The manager's local network is his/her political system. Key concepts are social networks, power and influence, interests, and dominant coalitions. The key processes are conflict management, negotiation, and forming and dissolving relationships. The manager's environment is his/her stakeholder network. His/her role as a leader is forging coalitions, identifying and leveraging interests, and negotiating. The stimuli for change in the network are shifts in dominant coalitions and shifts in the power of stakeholders. The largest barriers to change are "entrenched interests."

To navigate the political landscape, the first step is to identify groups, leaders, and bridges within the network. The next step, developing a stakeholders' map of connections and interests, is not easy, but it is the starting point for determining how the network will evolve, tear, and heal. That determination involves assessing (1) the role of propinquity in the propagation of information, (2) the nature of relationships (are they hierarchical or peer), and (3) the sociology of conflict resolution across the many links. Because not all conflict is due to miscommunication, one must find (political) ways to manage it. Politics yields the most influence when it is performed invisibly and legitimately so as to provoke little resistance.

Axiom: No matter who screws up, it's the boss's fault.
Theorem: Power accrues to those who are central in the network.
Corollary: If you aspire to be the executive entrepreneur, you must make yourself aware of your entire network.

Measures of the Research Manager

2

Theorem: You can't be a great leader if no one wants to follow you.

ELM SPACE

In many organizations, the most frequent topic of managerial training focuses, to an overwhelming degree, on *leadership* to the exclusion of developing other competencies of the manager and executive. A typical motto used to justify this approach is that "Leadership is the heart and soul, the single most important ingredient, of management, because what you manage is people." The premise "what you manage is people" is at least a partial truth, but the conclusion is exaggerated if not a mild sophistry. On that basis, one might claim equal justification that a degree in clinical psychology is the heart and soul of management.

Unfortunately, such statements overstate the obvious; a leader with no management competencies is likely the first lemming over the edge of the cliff. It does take skilled management abilities to attract the best people and to coach, and then to develop, and retain them. Without effective management, sound decisions cannot be made in any business area be it product production, research focus, corporate finances, or human resources. Consequently, effective management of the research enterprise demands that its managers and executives develop and practice a full set of skills that are appropriate to their job responsibilities. This chapter aims to identify the requisite skill set appropriate to the job and to help the individuals develop their personal profile of skills consistent with their aspirations for career development.

To visualize that balance semi-quantitatively, the author has introduced a specialized skill space, the Executive–Leader–Manager (ELM) space. ELM space, as illustrated in Figure 2.1, is a three-dimensional space with an axis of Subject-Matter Expertise, a Conceptual Abilities axis, and an axis of Interpersonal Skills. In ELM space, these axes define three planes: the managerial plane, the executive plane, and the plane of the intellectual leader—now popularly called the "thought leader." One may envision a value from 0 to 10 to apply to each axis.

The aspirant for a job assignment should recognize that his assessment of the ideal locus in ELM space of skills required for success in the position could be quite different than the assessment of the hiring manager, the search committee for the job, or the senior executives of the enterprise. Indeed every significant stakeholder may view the locus somewhat differently.

The Conceptual and Interpersonal axes define the executive plane in the space. Competent executives should have a high score on both of those axes. Of course, they also should have sufficient subject-matter expertise to understand the technical challenges their organization faces. The tasks most often associated with the Conceptual dimension are formulating and enunciating organizational vision, establishing organizational values, planning the strategy of the enterprise, managing risks faced by the enterprise, and prioritizing resource allocation to programs and projects. The competencies and expectations associated with this axis include integrity, courage, credibility, professional status, and social skills. Conceptual acumen is critical to the executive's ability to illuminate the vision and value of the enterprise and to inspire excellence in its workforce. The chief executive officer (CEO) of the enterprise has the ultimate responsibility to advance the enterprise toward organizational excellence by

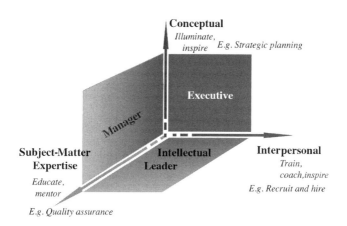

FIGURE 2.1   ELM space.

increasing the span of organizational accountability across all levels of the enterprise.

The space defined by the Conceptual and Expertise axes is primarily the domain of managers in the enterprise. Tasks associated with subject-matter expertise include managing resources and work schedules, evaluating and managing team performance, controlling and mitigating operational risks, and ensuring technical quality. In research organizations, the competencies and expectations of managers are the ability to make scientific and/or technical judgments, market research programs, make effective presentations, publish in professional journals, provide professional service, and gain professional status in their relevant research communities. Subject-matter expertise is critical to the managers' ability to educate and mentor their subordinates.

The Expertise and Interpersonal axes define the plane of the intellectual (or thought) leader and the organizational "guru." Tasks associated with the Interpersonal axis are coaching, providing expert advice, communication (including marketing), recruiting, developing and retaining staff, conflict management, and managing organizational change. The associated competencies or expectations are the projection of institutional values, ethical action, and timeliness. Quintessential activities are training, coaching, and inspiring other employees.

What is your ELM profile as seen by various members of your constituency? Sampled over a restricted set of stakeholders, such a multi-viewpoint perspective is the concept behind the 360° evaluation now popular in many organizations.

Topic for discussion or introspection: (1) Define a meaning to a few levels of scoring for each of the axes of ELM space. (2) Locate the last three heads of government in your country in ELM space. (3) Locate your laboratory director general or university president in ELM space. (4) Are these different? And why? (5) What do you see as your career trajectory in ELM space?

In research organizations, not all leaders are managers, and not all managers are leaders. Some managerial assignments primarily require technocratic skills. Projects offer a distinct case in point; the *project manager* is to deliver a well-defined task within a fixed time and within a fixed budget. The human side of personnel management becomes subordinate to keeping the project "on track." Not all talented researchers in the enterprise are well suited to assignments as project managers, yet projects large and small are the lifeblood of organizations that create and operate major research infrastructures. The task of the executives in the organization is to balance the discovery activities of the organization with its project execution activities and its obligations to operate and maintain the exiting research infrastructures of the organization.

Whether one aspires to be a "guru" or a top executive, certain personal characteristics are crucial to projecting *command presence* to one's colleagues and employees: courage, decisiveness, confidence, self-discipline, the ability to identify problems, and the willingness to devote most of his/her time to planning and controlling.

> **Theorem:** How you are perceived will influence how people hear what you say. *"People hear what they see."—Bobby Darren*

LEADING CREATIVE EMPLOYEES

Research into approaches to leading creative employees has followed three primary lines: (1) Personality research or "Who is the creative person"; (2) Cognitive research or the analysis of thinking patterns; (3) Exploring the logic of creative ideas; that is, reconstructing the objective logic of the ideas themselves. However, none of these inquiries address the challenges that the research manager must face every day.

Creative employees want the time and the freedom to exercise their creativity. Most recognize that they must also accept other assignments that occupy a large fraction of their time "to pay their dues." Unfortunately, economic realities and lack of sufficient staff can make it difficult to prevent obligatory tasks from occupying the entire day, every day, of every employee. The manager must be continually aware of this issue and find means and funding to preserve time for creativity. As a leader of creative employees, the manager's goal is to stimulate creativity by selecting the right staff, maintaining a stimulating environment, not overcommitting one's staff, and adopting an organizational structure (and compensation structure, if possible) that rewards creativity.

It is difficult to hire only geniuses because not that many truly great performers exist. Even if they did exist, no organization has that many top slots available. Nonetheless, when top talent does become available, it is worth a manager's effort to make the acquisition. But for the most part, successful, large research enterprises build their science and operations programs on highly reliable, good to very good scientists, engineers, and technical staff, plus some outstanding thought leaders and highly effective managers. From there, the manager must employ appropriate work processes to get outstanding deliverables from the team.

Creative scientific environments must recognize creativity *at the individual level*, because doing so only at the group level promotes "free-riding." Therefore, effective research managers need to establish incentives for individual creativity while also recognizing both the strengths and limitations of group thinking and team efforts. The strength of the team derives from the application of diverse skills and insights that can effectively amplify and expand seed ideas; in contrast, the limitations of group thinking are peer pressures for intellectual conformity. Overcoming these limitations requires the manager's attention to eliminate and avoid structural checks on the exchange of ideas or what is often called "pluralistic ignorance." The manager should also be aware that authority structures in themselves can stifle creativity. In the Navy, one might hear an officer say, "I thought that is what the Captain wanted me to do." Bosses can and often do "strongly suggest" conformity by their actions. For example, first President Trump tweets, then he publicly polls his cabinet one by one.

A particularly corrosive atmosphere is created when managers either suppress ideas or take credit for the ideas of subordinates. In such situations, creative thinking is punished; implying that "creativity is not in your job description" can quickly transform a strong team spirit into a herd mentality.

Several approaches can mitigate these negative influences in any group exercises. (1) Employ so-called anti-group ware; that is, anonymize inputs or use private signaling devices or polling techniques. In brainstorming sessions or in the group's planning retreats, set effective ground rules that encourage hearing all suggestions from anyone before any polling is done. Most importantly, in the discussion stage, the leader should listen to others rather than speak to the group. All these techniques work to deliver far better results in a trusting environment.

THEORIES OF LEADERSHIP

Caesar's boys fought for Rome, but most of all they fought for Caesar.

A quick search on the Internet[1] will uncover for the reader multiple models of leadership, both psychological and pragmatic. Many models try to answer the question, "Are leaders born or made?" Like most theories, the majority of models have some seed of realism at their heart, yet they go on to project a static—and often exaggerated—picture of what leaders are and how they function. Still it is useful for (aspiring) managers to be acquainted with the basic features of several popular models to highlight insights that might be relevant to their situation as well as to their personality.

---

[1] The list of suggested reading omits citing websites as many have little permanence on the web; therefore, the reader is urged to use his or her favorite search engine to find current live links.

The Vroom–Yetton Leadership Styles (VYLS) distinguish five degrees of authoritarian behavior of managers in making decisions. The most authoritarian manager makes decisions himself with whatever the information is on hand. As one moves away from authoritarian behavior, the group is asked to contribute increasing amounts of input in forms ranging from requests to gather additional information all the way to having the group making the decision with the leader acting only as the chair of the group discussion. This categorization may be interesting to some, but it does little to suggest which procedure may be most effective in which circumstances.

The Path–Goal theory of Evans and House tries to address that deficiency by advocating that the leader adopt a style that varies in response to the situation to affect the performance, satisfaction, and motivation of the subordinates. As in the VYLS approach, the styles range from "directive" to "motivational" in which the leader encourages the group to resolve the issue in conformity with the leader's challenging goals.

Likewise the Blanchard situational model breaks the action of the leader into "directive behavior" (directing and coaching) and "supportive behavior" (supporting and delegating). The rhetoric used by Blanchard implies his favoring the supportive category. The situational part of his model comes when Blanchard draws a *situational grid* plotting the degree of the leader's directive behavior against the leader's degree of supportive behavior. The model posits that the leader's directive behavior is roughly the inverse of the level of commitment of subordinates, while the leader's supportive behavior is roughly the inverse of the competence of the subordinates. Blanchard's situational grid is divided into four quadrants: the upper left quadrant calls for supporting behavior by the leader; the lower left calls for delegating behavior. These apply when employee competence is high. When employee competence is low, the upper right calls for coaching from the leader, while the lower right calls for directing behavior. If one ever watches professional sports, one will instantly note a deficiency of this model. Even very talented, committed players benefit greatly from excellent coaching. The Blanchard grid models have been subjected to parodies in the management literature; one such book suggests that a mnemonic for the lower right corner (low commitment/low competence employees) is that "one can't put a good edge on bad steel."

The Blake and Mouton managerial grid is similar to the Blanchard grid except that it plots "concern for people" against "concern for task." Again its authors break the space into four quadrants. Leadership that is concerned with neither people nor task is called impoverished while leadership with a high concern for both is "team leadership." The strong value judgments of the authors are completely summarized in their rhetoric of naming the styles of leadership. In short, Blake and Mouton tell one what to be, not how or when.

A fascinating description of styles is offered by the "Four Frameworks for Leadership" by Bolman and Deal. These authors suggest that approaches of four distinct frameworks—Structural, Human Resource, Political, and Symbolic—are useful in managing organizations. They suggest when each style is most effective and also give contraindications for each. Some managers are driven by their personalities to use one style to the predominance over the rest. The reader is encouraged to search the Internet for current references.

While the reader may find insights in each model, they all share a common deficiency. They are quasi-static in the sense that they do not reflect the dynamics and the degrees of connection and relationship between the manager and the staff. To be effective, the manager should develop a visceral understanding of the temporal dynamics of control. Those familiar with being the skipper of a small sailboat or being a referee on a football pitch will know that the style of sailing the boat (or refereeing the teams) is both situational and dynamic. Noting that freedom to be innovative and creative is highly regarded in research organizations, the morale of the team will be highest when the manager can let the team "play on" with minimal interference but with constant vigilance lest the game get out of control as depicted in Figure 2.2.

The skipper will recognize the level of direct control as being the pressure that his or her hand must maintain on the tiller. The referee will interpret the level as how tightly s/he calls the game. The referee maintains control of the game when s/he wins and maintains the respect of both teams by being in the right position to control the game and by knowing when to stop play and when to allow the athletes to play on.

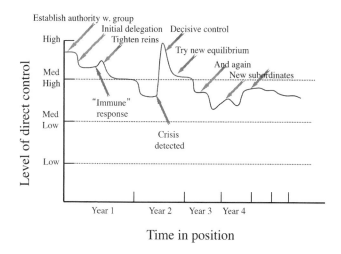

FIGURE 2.2   The ups and downs of dynamic team leadership.

All the theoretical models miss the fact that leadership depends on the *personal dynamics* of the manager with members of the team. A leader must be able to say to a staff member, "I need *you* to do this for *me*" and to know that the staff member will make the effort requested and required by the situation. Caesar's boys fought for Caesar.

Topic for discussion or introspection: What is the difference between manipulating people and inspiring them as a leader? Is this distinction different in a laboratory versus university versus corporate environment? How would you strike a balance in your job?

Theorem: "It's always personal" (The Godfather, Part 1).

MANAGING ORGANIZATIONAL CHANGE

Change happens every day; it's the magnitude that varies. What causes unplanned organizational change? (1) New people arrive; some staff members leave; (2) There are new assignments; staff become more proficient in their assignments; (3) New funding arrives—Inflation eats into purchasing power of present funding; (4) New projects/programs excite the staff—Staff become bored with the same old thing; (5) New customers and old customers are represented by new faces.

The integrated effects of the changing conditions can be considerable. In the case of change that is initiated from the bottom up, the change can be emergent, continuous, and incremental. To the extent that the direction of bottom-up change is radical, the top-down response of managers of the organization is likely to be reactionary and evolutional.

Organizational change can also be planned and top down. What causes top-down, planned organizational change? New management, crisis resolution, reorganization/restructuring, or major new business initiatives. In these cases, the intended effects are large and can be rapid, planned, discontinuous, and radical. Middle management can also initiate planned, downward change. At all levels of management, the dynamics of change can easily produce organizational sheer that is felt most acutely by middle management. Top management should be aware of and have plans to react to bottom-up resistance from the "immune system" and organizational inertia of the enterprise that seeks to mitigate the speed and the effects of change on multiple levels.

Managing through change requires engagement at every level:

1. At the perceptional level are (1) Staff members' fears (of losing power, value, status, independence, employment, benefits, and even their job); (2) Staff resentments; (3) Performance anxieties. Managing these potential perceptions requires understanding how *others* see the situation.
2. Intellectual reactions include the following: (1) Advertised benefits of change aren't convincing and (2) Rebellion: a time of change is an opportunity to challenge authority by convincing others that the status quo is not working. A manager must be able to explain the logic of the change agenda.
3. Operational reactions: The time required to acquire the new skills and the time required to learn new processes are not worth the benefit. When change comes from above, the manager should lead by example. As a middle manager, you must support your superiors.

Before initiating change, a manager needs to make a correct diagnosis through a careful analysis of the situation. It helps to have the employees in the organization (or sub-unit thereof) involved in the situational analysis and the characterization of the trade-offs of alternative courses of action. With those prerequisites, the manager can set realistic goals. In the university setting, "consultation" with other faculty members and senior administration officials to "socialize" the need for change is especially important. To manage implementation of the specified change, the managers should first assure sufficient staff buy-in and the availability of sufficient resources to carry out the change.

In other words, identify what *prize* is gained by the change and what *price* the enterprise must pay. Be realistic about evaluating the chances of success. Ask "Is this for you?" Learn how others see you and what you represent, and then ask yourself "Are your ends impeding the projected benefits of change?" Then proceed decisively.

One of employees' most common complaints about managers is indecisiveness. As one will never have 100% of relevant information, any manager/leader must make the most of whatever information is at hand. In the end, being decisive requires setting a timetable for deciding and making that known to all key stakeholders. At an early stage, managers should schedule consultations with relevant stakeholders and encourage further input up to the announced deadline. Then the manager should make a decision in accordance with the schedule as announced. When a manager expects decisions to be made by subordinates (whether an individual or committee), s/he should be clear about when the decisions must be made. It is quite effective to announce that if the decision is not forthcoming at the prescribed time, then the manager will make the decision for the subordinates. Then next time around, it is more likely they will meet the specified decision deadline.

> **Theorem:** You will undermine your authority as a leader, if you do not know and exercise your rights as a manager.
> **Corollary:** Articulate your rights as part of your expectations of your employees.

As a research manager you have a right to require excellence; be consistent about your standards. You can expect your employees to conduct their research commensurate with professional ethics and the guiding principles of the enterprise. The nature of both professional research ethics and government-imposed regulations governing the use of its funds should be made clear to all the employees. You have a right to require that employees follow your instructions (unless they are unsafe or illegal); failure to do so is insubordination and is a cause for disciplinary action even if the issue is substantively trivial. You may add or delete job duties for employees who are not represented by a union. However, changes in work rules for represented workers must be approved by the union representative (shop steward). Finally, you have a right to expect good conduct from employees. That includes respectful behavior toward all in the workplace.

Likewise, employees should understand the rules of engagement. (1) They must tell their employer what they learn concerning the organization's business; the enterprise has a right to expect their loyalty. (2) Employees must not reveal confidential or competition-sensitive business information. (3) Employees may not compete with their employer and must comply with organizational policy concerning any outside employment. (4) Employees are expected to give a good day's work for a day's pay and maintain an adequate attendance record. This last expectation brings to mind the oft-repeated quip from workers in Soviet Russia, "We pretend to work, and they pretend to pay us."

The Research Environment

3

WHAT IS RESEARCH?

In its broadest organizational context, research connotes the search for previously unknown information or the systematic study of methods and information, both descriptive and quantitative, that are not known to the enterprise or more generally are not openly known at all. More specifically in the world of science, engineering, and technology, research includes a very wide variety of studies, both theoretical and experimental, that must precede the development, production, and deployment (or marketing and sales) of useful materials, methods, devices, or technical systems that are of economic value in a broad sense. The U.S. National Science Foundation (NSF) has produced a compendium[1] of definitions for research and development (R&D) that are used in the United States and Europe. Throughout the sources cited by the NSF, three general categories of activities are distinguished: *basic research, applied research,* and *experimental development.*

Basic research connotes a systematic *study* to acquire new knowledge of the fundamental aspects of phenomena without regard to any specific applications toward deployable methods, processes, or products. Basic research enters the realm of *applied research* when the information it generates becomes necessary for determining the means of meeting a recognized and specific need or application. Finally, *experimental development* is the systematic use of the knowledge gained from previous research to produce useful and testable methods, materials, products, or systems. Experimental development generally includes the design, development, and testing of prototypes and pilot processes.

---

[1] "Definitions of Research and Development: An Annotated Compilation of Official Sources," https://www.nsf.gov/statistics/randdef/rd-definitions.pdf (March, 2018).

These definitions are most frequently applied by governmental funding agencies and philanthropic foundations. Some agencies further refine these categories during the allocation of their R&D budgets. For example, the U.S. Department of Defense distinguishes seven levels of R&D, giving Test and Evaluation as its own separate category and referring to the entire sequence of activities as RDT&E. An alternative categorization of R&D activities may be more valuable for use by investors and research enterprises, which are investing their own internal funds in R&D activities. This breakdown distinguishes the levels of technical risk and the time needed to bring an application of applied research to a marketable product. In that nomenclature, investors (either internal or external) expect *short-range (incremental) research* to have a success rate greater than 50% and a time to first product of at most 2 years. Such characteristics imply that all the fundamentals and their application to a marketable application are already known and tested. This category is, in fact, quite close to the end phases of experimental development plus the test and evaluation stage; it is driven by a specific product line and its detailed characteristics.

Research with a time horizon of 2–5 years involves the late stages of applied research, and the time duration of *mid-range research* can depend strongly on the competencies of the enterprise and the efforts made to protect the dissemination of research results at too early a time—thereby giving up some competitive advantage. The probability of economic payoff of mid-range research is in the range of 20%–50%. With appropriate protection via patents or trade secrecy, such applied research plus experimental development can provide a more durable competitive advantage to the enterprise than short-range research, which may lure strong competitors into the market.

Not surprisingly, *long-range or fundamental research* has the potential to deliver a very large competitive advantage to the enterprise based on observing, characterizing, and controlling breakthrough phenomena; however, its risks are large and are difficult to assess at the outset. A time horizon of 5–10 years can be expected, although some research programs such as those aimed at developing quantum computers may stretch to 20 years or more. Certainly, for basic research undertaken by governments, the time horizon is typically greater than 10 years with little guarantee of direct economic payoff.

Topics for discussion or introspection: (1) Is managing a research enterprise different from managing a manufacturing company? A retailer? A law or consulting firm? Explain your answer. (2) How does your organization decide which research to conduct?

TRENDS IN RESEARCH FUNDING

In most countries, the funding for R&D since 2006 has been relatively flat except for a marked dip during the global recession of 2008–2010. According to the European Commission,[2] from 2006 to 2016, the combined national spending of EU countries on R&D has been roughly 2% of the EU Gross Domestic Product (GDP), although spending on major research infrastructures has increased dramatically since 2012. The fastest rate of growth in Europe has been 3.4% in Switzerland dominated by in-house research by major pharmaceutical companies. Comparable statistics for the U.S. (excluding capital expenditures) and Japan are ~2.7% and ~3.3%, respectively. In contrast, the expenditure in China has risen steadily in this period from 1.6% to nearly 2.1%, and it continues to increase.

In terms of actual currency, the NSF estimated the total global expenditure on R&D in 2015 as ~$1.92T (in terms of purchasing power parity[3]). That year, the total expenditure of the United States on all types of R&D was $499B. Funding in Europe is roughly 80% of that of the United States. Of the total funding in the U.S., the NSF classifies one-sixth as basic research, one-sixth as applied research, and two-thirds as engineering development. Most of the engineering development is devoted to in-house-funded, industrial R&D. Taking all economic sectors into account, private sector–funded research in the United States is almost triple that of government-funded R&D.

The higher s/he rises in the organization of a research enterprise the more important it is for the manager to understand the market (or funding) opportunities and the trends in the science or technology sectors that are relevant to the enterprise. The partners in a small start-up should have such awareness before they take the plunge of establishing their business. Certainly they can expect any investors into their venture to evaluate market conditions and near to mid-term technology trends.

---

[2] "Eurostat—Statistics explained" https://ec.europa.eu/eurostat/statistics-explained/index.php/R_%26_D_expenditure.

[3] Purchasing Power Parity compares the value of national currencies based on a "basket of goods" approach.

THE R&D LIFE CYCLE

Whether one is considering the financial cycle from the investment in research to the commercialization and sales of a single product line or the cash flows of a technological start-up enterprise, the typical financial expenditures through the various phases of research, engineering development, and commercialization or deployment can be visualized schematically as depicted in Figure 3.1.

The classic S-curve (in gray dots) indicates how the performance of a new technology improves with time and with the associated effort invested into it. The dashed, black curve that indicates the level of R&D expenditure per unit time is approximately the derivative of the gray curve.

First sales of the product to early adopters begin early in the commercialization phase. Until revenue from sales surpasses the expenditures for the R&D cycle, the annual cash flow for the business or product line will be negative. The integrated financial balance for bringing the product to market (or deployment) will remain negative until the income from sales of the new product line grows sufficiently along its own S-shaped curve that is roughly proportional to the S-curve of performance, and that indicates sufficient market penetration (or technology adoption in the form of major research infrastructures).

In short, the business plan of the enterprise must tolerate the years of negative cash flow in the effort to carry a new product from the research department to sales. However, the financial backers of the research efforts generally have limited tolerance for negative cash flow, especially when it is not partially offset by any income. Therefore, senior executives of the

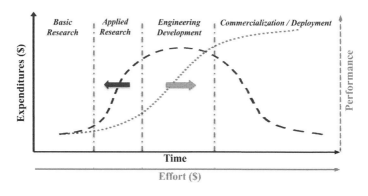

FIGURE 3.1   The R&D life cycle.

enterprise must be able to make realistic forecasts of the time scales of applied research, engineering development, and commercialization in their business. In addition, top management must communicate and convince its financial backers of the strong competitive advantage that the enterprise will gain in the marketplace.

The research managers in a national laboratory, university, or other not-for-profit institution may object that the issue of cash flow does not apply to them or their organization. To the contrary, the analogues do exist and are commonly assessed by the funding agencies. Moreover, the level of sustained or growing funding of research programs is vital to these not-for-profit organizations to generate overhead funds that can pay for the upkeep and reno-vation of the infrastructure of the organization. The analogues to profit may be converting a short-term research program into the construction of a major infrastructure, securing a lucrative technology-transfer opportunity (either licensing or sale of patents or spinning off a commercial enterprise), leading large multi-year programs or high-technology collaborations, or becoming the lead institution in the funding agency's new, high-impact scientific initiative. Although financial profit does not apply, funding agencies do and increasingly must create other metrics to keep score of scientific and technological success. The savvy research manager learns what these metrics are and leads the enter-prise to deliver a winning score.

YOUR WORLD AS AN R&D MANAGER

For executives and managers to drive their enterprise to function effectively in the research environment, they need to assess the position of their enterprise in that environment. To begin, they should understand its dimensions qualita-tively (or even better semi-quantitatively) as they apply to their organization. One can visualize the enterprise as existing in the three-dimensional *research management space* of Figure 3.2a, the dimensions of which are Mission, Competencies, and Market.

1. *Mission* describes what the organization is supposed to deliver. This description is likely to be quite different for single-purpose organi-zations versus multi-purpose laboratories or corporations.
2. *Competencies* describe what the organization is actually capable of delivering. The competencies are shaped by technical breadth and depth of the staff and by active strategic partnerships.

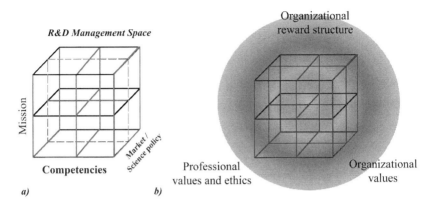

FIGURE 3.2   Research management space.

3. *Market/science policy* answers the questions, "Who cares about what the organization promises to deliver? Will anyone pay for the research that it would like to perform? How will that research influence the demand environment?"

The CEO of the enterprise will see the management space from a top-down perspective. In contrast, the non-managerial employees of the organization have a bottom-up view of this world. Any manager, regardless of position in the organizational hierarchy, should realize that these two perspectives are almost certainly likely to be quite different. S/he should understand and will have to manage these differences in perception.

Embedding the organization in a *value space* recognizes the human side of the enterprise (Figure 3.2b). The value space is described by three principal axes: (1) the *organizational reward structure*, (2) the *organizational values*, and (3) the relevant *professional values and ethics*.

> *Topic for discussion or introspection*: Discuss your organization with respect to Mission, Competencies, and Market/Policy.

The values embedded in the *organizational reward structure* are highly dependent on organizational (and national) culture. Obvious aspects of the reward structure are respect associated with position in the hierarchy of the organization and its implications for promotions, professional growth opportunities, and job titles. Rewards also include recognition via internal and external visibility (participation in international conferences, for example) and reputation, social recognition, bonuses, and recognition awards. A boss sincerely saying "thank you" is also a valued (and underused) reward.

Security is the second aspect of the *organizational reward structure*. Compensation levels, job security (and turnover rate), group harmony, and a sense of belonging or affiliation all characterize security as an organization reward. Likewise professional stimulation is a third avenue of reward. Examples are providing new and exciting job opportunities, gaining mastery in a new domain, intellectual independence, permission for risk-taking, and access to internal R&D funds.

Ask yourself what your organization values. As the leader of the enterprise or one of its sub-units, one of your obligations is to act in a manner that models the values of the organization to your subordinates. The higher your position on the organization chart, the greater is your obligation.

Commercial research organizations value profit (or the generation of overhead funds in a not-for-profit enterprise), technical innovation (for example, obtaining patents and winning R&D 100 awards), entrepreneurship, risk-taking, delivering on promises to stakeholders, recognition for best business practices, professional awards, intellectual leadership both nationally and internationally, service to the nation, and service to the local community.

Professional values include publications and productivity, service to the profession through professional organizations, service on peer review committees and peer reviewing articles for professional journals, and public outreach. Relevant ethical values include integrity, collegiality, and confidentiality. As a manager, you should understand and model the professional values that drive your organization's behavior and that are most important to your creative employees.

Topic for discussion or introspection: (a) With respect to organizational values, how do corporations differ from government laboratories? (b) From the point of view of motivation in R&D organizations, how do research scientists differ from engineers? From technical staff? From administrative staff? How might your answers differ for a multi-cultural work environment?

Strategy, Forecasting, and Technological Risk

4

If you don't know where you're going, you won't know when you get there.

In the context of a research organization, the goal of strategy is to achieve superior long-term return on investment (either internal or external) by creating sufficient economic, scientific, or technological value for which the customers (for example, one's research agencies) are willing to pay for deploying the products of the enterprise and for growing the competencies of the enterprise.

CHARACTERISTICS OF A BUSINESS STRATEGY

A *strategy* is a distinct system of activities that guides the enterprise in pursuing choices amongst competing options that create and capture unique economic, scientific, or technological value rather than merely improving processes or operational effectiveness. Strategy should be deliberate, planned, and pursued to realize the intended goals of the enterprise. While process improvement is undoubtedly important to advancing the enterprise, it is an aspect of effective tactics of business operations rather than strategy.

Following a deliberate strategy allows the organization to transform its articulated goals into action, to focus the direction of its changes, and to make choices. It enables the enterprise to terminate fruitless activities and to change course when necessary. Without an articulated strategy, it is very hard to kill projects. Not surprisingly, effective strategies change over the research and development life cycle. To summarize, sound strategy makes a better future more probable. These thoughts are illustrated schematically in Figure 4.1.

In the absence of a clear strategy, management may simply drive the enterprise ahead in time guided only by wishful thinking regardless of the evolution of either internal pressures or external circumstances. Such an approach is most likely to lead to an inevitable decline in the overall performance of the organization. A somewhat better approach for management is to react to externalities in a manner that is designed, or at least intended, to mitigate negative influences, and to chase after perceived opportunities.

Why might such reactive management breed failure? First, it is *ipso facto* based on trailing indicators or, to use an oft-used expression, "dressing for yesterday's weather." Second, it is highly susceptible to suffer from an unrealistic view of present conditions and from ignorance of the external environment, that is, growing strength of competitors, new regulatory barriers, and other business constraints. Reactive management generally suffers from a poor understanding of institutional dynamics within the enterprise. These deficiencies can lead to a poor evaluation of risk elements, which results in a dim, unrealistic vision for the future. In contrast, a sound strategy manages risks to the enterprise based upon realistic evaluations of internal and external conditions and on technological forecasts, both of which are updated at regularly planned periods.

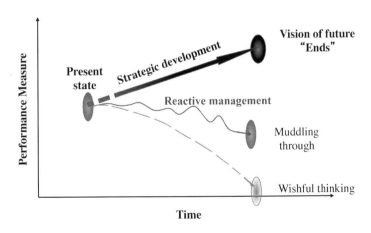

FIGURE 4.1   Why have a strategy? Answer: To build for a better future.

ENTERPRISE RISK MANAGEMENT

A fundamental task of all levels of management is to mitigate risks to the organization within their span of control. For that reason, the Board of Directors holds the chief executive officer (CEO) (or Laboratory Director General) responsible as the chief risk officer of the enterprise regardless of what title regarding risk the CEO may bestow on an underling. The fundamental reasons for managers to implement a process of risk management and mitigation are (1) to minimize negative impacts on the organization and (2) to provide a sound basis in decision-making.

Risk management is the process of reducing risk to an acceptable level, that is, to a non-catastrophic level that is readily amenable to internal correction and for which the enterprise and its management are safe from charges of negligence. The steps involved in the risk management process are (1) identifying the risks, (2) characterizing the risks, (3) assessing the consequences of risks, (4) mitigating risks, and (5) reassessing the presence of remaining risks and their consequences. For this purpose, the generally accepted quantification of risk of an event is the expected value of its cost as given by the formula

$$\langle \text{Risk} \rangle = \text{Probability of the event} \times \text{Cost of damages from the event,}$$

where Cost is the maximum credible loss from the event or class of events. Enterprise managers are likely to be found negligent in tort or criminal legal proceedings if the amount that they invested to mitigate the risk is less than the expected value, $\langle \text{Risk} \rangle$. The flip side to risk is unforeseen benefit; in that case, the cost is assigned a negative value.

The management characteristics that optimize strategy and risk management are anticipation and adaptation. As applied to research and innovation in science and technology, risk management refers to a system of cost-effective countermeasures deployed in advance that control the uncertainties associated with a potential negative impact on achieving the research investment goals of the organization. The mechanisms used to apply these countermeasures to research processes should be based on an analysis of their feasibility and effectiveness. The resources expended in implementing the process in organizations should be based on a cost-benefit analysis that may be either qualitative or quantitative. The aim of the analysis is to demonstrate that costs of implementing controls can be justified by the reduction in the level of risk. Proper documentation of the risk management process is essential to protect the enterprise and its managers from tort liability or even criminal charges.

Understandably, research managers act out of optimism concerning the results, impact, time scales, and costs of the research efforts of their teams. Yet these quantities carry considerable uncertainty because by its very nature research involves that which is unknown. This uncertainty is *technological risk*. The management of technological risk is essential to the successful completion of science and technology programs and projects. Again cost-benefit analysis plays a central role; the steps of a cost-benefit analysis are the following:

1. Determine the impact of implementing new or enhanced controls.
2. Determine the impact of not implementing new or enhanced controls.
3. Estimate the costs of the implementation. These costs may include additional research, hardware purchases, costs associated with additional procedures, and costs of hiring additional personnel.
4. Assess whether the controls are worth the cost.

The U.S. government publication NIST Special Publication 300–30 states the rationale succinctly. "Just as there is a cost for implementing a needed control, there is a cost for not implementing it."

Successfully managing risks in research organizations demands recognizing all forms of technological risk. One may distinguish five general categories: technology factors, policy factors, market (or social) factors, legal factors, and moral factors.

Policy factors include (1) the over- or under-estimation of economic feasibility, (2) deficient forecasting of trends in science and technology, and (3) the absence of reasonable criteria governing resource allocation.

Market and social factors can include (1) the release of the new product or service ahead of societal demand, (2) conflict among new and competitive products or services, and (3) the lack of social consensus concerning the use and utility of new science/technologies, for example, public fears concerning nuclear power, irradiated foods, and genetically modified organisms.

Legal factors: Conflict between the new product or service and existing regulations, conflicting international standards.

Moral factors: Inappropriate use of research funds, leaking of research secrets or other confidential information, outflow of the research workforce, presentation of exaggerated or false research achievements.

Managing these technological risks requires well-formulated (and documented) *contingency plans* that are weighted according to a cost-benefit analysis. Having contingency plans implies having forecasts of what will or might happen. The prudent management team makes these assessments frequently but at least annually.

TECHNOLOGICAL FORECASTING

A major challenge in managing research organizations is managing the uncertainty that accompanies probing into the unknown; similar considerations apply to project management. Generally, forecasting deals with three classes of events that inject uncertainty into a program or project: First are known, expected effects that derive from the emergent phenomena in complex systems. Second, recognized uncertainties, referred to as "known unknowns," can introduce effects of indeterminate magnitude on critical decisions; mitigating these effects requires constant vigilance, identification and sensitivity analysis of the effects and their impact, and a flexible research strategy able to mitigate the adverse impact. Third and finally, the irreducible uncertainties of "unknown unknowns" require (1) constant monitoring of all activities that may have a negative impact on key decision points and (2) policies that can incorporate lessons learned and then adapt to changing circumstances. The goal of technology forecasting is to spawn the corrective actions required for the enterprise to stay on course with its strategic plans.

Whether one must mitigate uncertainties from known effects or known unknowns, an *integrated assessment* makes best use of the limited initial information *before* decisions are enacted, taking into account the irreversibility of public policies (and business strategies) in the face of subsequently revealed information. Integrated assessment emphasizes the formal analysis of costs, benefits, and technical risks. Admittedly, anticipation of potential risks is difficult, and early warning signs may not be taken seriously, but top management must establish procedures to identify and document both major risks and their plans for mitigation *before* the risks are realized. Suggestions for anticipatory risk management include (1) improving and certifying the quality of data, analytic methods and models as legitimate and credible; (2) strengthening organizational structures for credibly assessing early warnings; and (3) enhancing transparency in the selection of leading indicators and models.

Large projects often account for both known effects and known unknowns by formulating a database of risks along with mitigating actions. This list, called a risk registry, also notes whether there is a defined time in the project (or program) at which the risk goes to zero because the activity or related sub-system has been completed. The registry is updated frequently, often on a monthly basis. The registry allows the principal manager to manage any contingency funds under his/her control to assure that they remain large enough to mitigate remaining risks.

After-the-fact integrated reassessment of risk mitigation strategies should use all information revealed plus an analysis of the relevant networks of the organization to improve the quality and timeliness of organizational adaptation to potential risks. As an organization's policies tend to lock in, management needs to look for opportunities to reduce interests in the status quo by strengthening incentives for surfacing and using information that reveals risks bred by earlier decisions. Early anticipatory analysis and prompt corrective action can keep the enterprise on its strategic course or even allow it to complete the course faster, better, and cheaper.

Techniques used to forecast the future of scientific understanding and technological progress range from little better than hunches by executives and their consultants to complicated mathematical analyses. They may be grouped under a few main categories: (1) judgment methods, (2) counting methods,[1] (3) times series methods, and (4) casual methods.

Among the judgment methods are naïve extrapolation from the recent past and consultation with relevant geniuses and panels of experts. Figure 4.2 shows an example of an historical trend of performing nuclear and particle physics with beams of protons hitting stationary (fixed) targets in the laboratory. The terms in italics are the important discoveries in physics made with

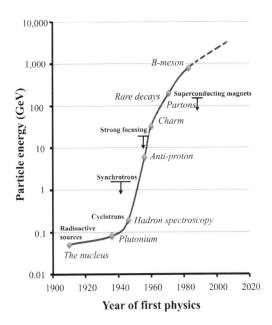

FIGURE 4.2   The fixed target paradigm of high-energy physics experiments.

---

[1] Counting methods generally apply to forecasts of market evolution. They may take the form of market testing and customer surveys.

the high-energy beams. Terms in regular font are the technologies used to generate the beams of ever-higher energy.

One can note several features from this example. First, the trend curve is not a straight line; naïve straight-line extrapolation would be incorrect. Instead the curve displays an S-shape that is common for developing technologies. Second, there are no more important discoveries in this field of physics directly using high-energy beams hitting fixed targets. Instead as the S-curve[2] was starting to turn over, a new technological paradigm, particle colliders, came to the fore. Under the old paradigm, the energy range of exploration grew as the square root of particle energy; with colliders, the energy reach grows linearly with the beam energy. This type of behavior is typical of technological advances. Examining relevant accelerator technologies from high-voltage DC generators to radio frequency linear accelerators (RF-linacs) to cyclotrons to synchrotrons more deeply, one finds that the S-curve in Figure 4.2 is actually the outer envelope of the S-curves of the several specific technologies. As given technologies advance in performance with diminishing returns along an S-curve, innovators invent new technologies to jump to a new S-curve that is just at its beginning of growth. The historical trend displayed in Figure 4.2 does not derive from the failure of synchrotron technology but from the underlying physics of the experimental paradigm. The economics of the fixed target approach scales unfavorably because the cost of the accelerator facility scales as the square of the energy reach (center of mass energy) of the experiment. In contrast, the new paradigm of colliding beams still uses proton synchrotron technology, but the collider paradigm allows the cost to increase slightly less than linearly with the energy reach. Accelerating beams to fixed targets remains important, but its use has become producing secondary beams of neutrons, neutrinos, and rare isotopes.

Another example of an historical account of progress is the oft-cited Moore's Law. In 1965, a co-founder of Intel Inc., Gordon Moore, noted that the density of transistors on integrated circuit chips was doubling every ~24 months. That trend has continued to persist, although the data set from 1970 to 2016 (showing a million-fold growth in density) actually shows a classic S-curve with a slowing of growth in roughly 2008. The underlying science that accounts for the slowing is that as the distance between neighboring transistors becomes less than ~15 nm, the mutual inductance of their respective circuits introduces crosstalk between the circuits. A technological impediment is that at such high circuit densities, the waste heat must be removed by even more aggressive cooling, including the immersion of the chips in cryogenic fluids. A further technological obstacle has been the need to develop radiation sources, optics, and metrology for lithography at the ~15 nm scale. The explanation for

---

[2] Note that the S-curve is just the time integral of the bell-shaped curve of technology innovation that was introduced in Chapter 3.

the trend of Moore's Law, despite the costs of driving the technology of chip manufacture, is a tightly coupled metric for integrated circuits: the number of calculations per second per $1,000. This metric reveals the powerful economic incentive that explains the extremely large expenditure of funds by the semi-conductor industry into advancing the technologies of CMOS (complementary metal–oxide semiconductor) chip manufacture. Interestingly the cost formulation of Moore's Law does not yet show a slowdown in growth.

The density of circuits is now near the level where crosstalk is inevitable; people often say that the era of CMOS chips on silicon is near its death. The actual response, driven by investment by large corporations with deep research pockets, is to develop a practical quantum computer that exploits the phenomenon of quantum entanglement to form networks of quantum bits (qubits). Just as in the case of accelerator technology for high-energy physics, one anticipates a jump in capability of chips per unit cost by 10^3–10^4 at an affordable cost. Whether quantum computers can deliver this promise for a universal, general purpose computer remains to be demonstrated.

A second judgment method is consulting relevant geniuses and/or panels of experts. The latter choice is most popularly used in consensus processes, in which opinions are solicited in a formal manner from a panel of experts and the synthesis of opinions yields the final forecast. One realization after consultation with experts is the Delphi technique of polling, collating, and refining independent forecasts based on multiple cycles of the process. A very common variant used in national laboratories and universities is the scientific (or technical) advisory committee or visiting committee. Such committees generally review the past and make extrapolations into the future to advise the senior management of the organization. Their output is often couched in terms of observations, findings (items of concern), and recommendations (advised actions). The manager should consider committee's advice seriously keeping the caveat in mind that advisory committees frequently issue what are essentially unfunded mandates; that is, the experts are responsible for neither cost nor schedule.

Another form of managing risks based on past experience is assigning budget and schedule contingencies in projects to mitigate uncertainties. Table 4.1 shows a practice that is commonly used in the U.S. to manage project risks. To use the table, the project manager first breaks down the work of the project into a large set of logically separated tasks (the work breakdown structure). The project team generates an estimated cost for each sub-task. Then a budgetary contingency for each sub-task is assigned according to the paradigm of the table. The total estimated cost is the sum of the baseline costs plus the sum of all contingencies.

Time-series methods using multivariate statistical regression aim to quantify past experience to insert some mathematical rigor into the process of trend

TABLE 4.1 Example of cost risks and contingencies in U.S. government technological projects

RISK FACTOR	TECHNICAL BASIS	DESIGN LEVEL	BASIS OF ESTIMATE	CONTINGENCY (%)
15	Designs beyond state-of-technology	Technical concept	Scaling relations	60
10	Considerable R&D needed to apply state-of-the-art	Technical concept	Analogy with existing designs	40
8	Some R&D required to apply state-of-the-art	Concept with sketches	Engineering estimate	35
7	Considerable extrapolation of existing designs	Concept with sketches	Engineering estimate	35
5	No significant advances over state-of-the-art technology	Partial drawing package	Vendor cost range	30
4	Minor modification of existing designs	Reviewable engineering	Vendor quote	25
3	Design with engineering drawing package	Full engineering drawing package	Vendor quote	20
2	Fully reviewed engineering, all materials acquired	Fully reviewed design	Firm vendor offer	15
1	Fabrication >50% complete	Design in fabrication	Firm, fixed priced contract	10
0	Commodity hardware	Off-the-shelf, firm price	Firm, fixed priced procurement	5

extrapolation; however, complicated mathematics that does not take causal factors (such as underlying science and economics) into account cannot produce trustworthy forecasts. Monte Carlo methods have proven to be very useful in complex projects such as the design of large detectors to estimate uncertainties in overall system performance. They are frequently used in conjunction with the risk registry to evaluate the adequacy of contingency funds held by the project manager as a percentage of cost to complete the project.

Cross-impact matrix methods aim to account for the probability that the occurrence of one event can influence the probability of other events. The method searches for the structural correlations within the system as a whole. The range of such mathematical methods and causal modeling is well beyond the scope of this chapter. However, the user must keep in mind that the complexity of the methods does not guarantee knowledge of the future.

Managers of large and small enterprises need documented plans to deal with uncertainties. These plans need frequent, at least annual, reassessment and revision. In any case, familiarity with historical trends and relevant technological surprises is an excellent starting point for managers at all levels. With this background, we are ready to make a plan.

Introduction to Strategic Planning

5

Theorem: A dog can't chase every car.

The research environment abounds in opportunities, far more than any organization can pursue. Moreover, for each opportunity pursued, the organization must spend discernable resources, none of which are infinite. Questions that follow from the theorem are simple: How does the dog decide which car to chase? And what would it do if it caught the car? If you are the top dog of the enterprise, you need to know which criteria you will use to decide these questions as well as what to do to maximize your chances of success.

PREPARING TO CRAFT THE CONTENT OF THE PLAN

Chapter 4 introduced the concept of strategy as the roadmap for building the enterprise toward a better future; the roadmap and its rationale were depicted schematically in Figure 4.1. This chapter develops the simple arrow of that figure into an actual plan that the organization can execute.

Good plans have several important characteristics. They are based on a realistic understanding of the conditions in the present research environment, including knowing the competition and potential allies, the barriers to success, and the constraints on the freedom of action of the organization. Good plans project a coherent strategic vision for the future that can guide the management and excite the employees. They reflect a thorough understanding of institutional dynamics that govern the behavior of personnel within the organization. They set forth clearly the management's understanding of the external environment such as regulations and public perceptions. To initiate the preparation of a sound strategic plan, the top executives need to launch a set of pre-planning activities. Their actions will generally include most, if not all, of the following activities:

1. The first and most important task is to define the objectives, boundary conditions, and the time horizon of the planning exercise.

2. In concert with senior managers, top management selects the members of the planning team consisting of a representative sample of managers and key employees throughout the organization. The considerations in selecting the team are the areas of control of the managers selected, their positions within the organizational hierarchy, and the range of knowledge and abilities that are required on the team to ensure insightful planning. A further option to be considered is whether junior staff members should be represented.

3. Writing a plan is a project. It has a beginning, an end, and a defined objective. As with all projects, a schedule and a budget must be set at the outset. The project team will generally need logistical support. In particular, management should select a facilitator for meetings of the planning group and designate scribes to record and collate the deliberations of the planning group.

4. Almost certainly the planning group would benefit from receiving briefings from subject matter experts. If the experts are from outside the organization, signing of non-disclosure agreements is a prudent precaution.

5. A sound plan is intended to be crucial to the future of the organization. Skimping on hiring an expert facilitator and providing favorable logistical support and sufficient expert advice is false economy. In addition, one or more administrative or junior-level staff should be appointed to be scribes.

PHASE 1—GATHERING AND ASSESSING INFORMATION

Good plans begin with gathering sufficient information upon which a sound understanding can develop. Analysis of those data by the planning group to create a picture of the present status of the enterprise can then follow.

Most of the work of the planning group will be done in a series of meetings, one of which should be a multi-day, offsite *planning retreat* with all team members in attendance. At the outset of the retreat, a facilitator (rather than a senior manager) describes what process will be used to make decisions during the retreat; possible options are full (unanimous) consensus, votes by the group (top management should define what constitutes winning a vote—a majority, a plurality, or a supermajority, etc.), or "call the boss for his/her opinion." The appointed scribes record the proceedings of the group, the decisions made, and any matters that are deferred for later discussion and resolution in follow-on meetings.

The facilitator reminds all of the Vision, Mission, and Guiding Principles of the enterprise (the foundational documents) and the planning assumptions for the exercise—as specified by top management—in their simplest terms. The group discusses and expands the assumptions into an analysis of the circumstances of the enterprise at the time of the retreat. The group should have relevant organizational data at hand, or it will need a resource assistant not in the group to gather and then transmit (often in encoded form) such data back to the retreat participants.

A tool frequently used to organize the gathering of information and to begin its digestion and analysis by the planners is the "*gap analysis.*" The gap analysis should uncover what is to be gained and what deficiencies are to be ameliorated by the execution of the new strategy.

In the gap analysis, the planning group assesses both the relevant Internals—**S**trengths and **W**eaknesses—and the Externals—**O**pportunities and **T**hreats to the organization (SWOT). It is vital that the identification of weakness and threats be given full voice; doing so is essential to a realistic plan.

The next step is to gain clear perspectives on the organization's objectives, that is, to identify what the organization might want to change. The outward categories considered for change include

a. *Financials*: Revenues, costs, productivity, and capital assets
b. *Customers*: Market segment (R&D sponsor), relationship and loyalty, product/service requirements, and customer satisfaction.

The enterprise should also consider what inward changes it might make:

c. *Internal processes* (how we work): R&D innovations, production attributes, and supply chain logistics
d. *Infrastructure*: Employee satisfaction and productivity, information systems, physical plant, organizational and control structure, and policy and procedures.

Topic for discussion or introspection: How do these considerations apply to the scientific or engineering research environment? Give examples.

TABLE 5.1   The gap analysis in the form of a balanced scorecard

	STRENGTHS	WEAKNESSES	OPPORTUNITIES	THREATS
Financials	Financial strengths	Financial weaknesses	Financial opportunities	Financial threats
Customers	Customer strengths	Customer weaknesses	Customer opportunities	Customer threats
Internal processes	Process strengths	Process weaknesses	Process opportunities	Process threats
Organization	Organizational strengths	Organizational weaknesses	Organizational opportunities	Organizational threats

The elements of SWOT and the aspects of the gap analysis can be summarized in a single mnemonic called the "balanced scorecard."[1] That formulation is depicted in Table 5.1.

The stated strategic foundations of the enterprise should shape the gap analysis. Besides the Mission and Vision statements, a research organization may have some selection rules to apply especially in deciding closely competing matters. One example of a guiding principle might be, "Outstanding, peer-reviewed science is the best foundation for new programs and projects." A priority selection rule could be, "Prefer one world-leading program to two follower programs." The strategic foundations are critical inputs by the top management and the governing board of the enterprise. Phase 1 of the retreat should result in a good picture of the strategic goals, the "ends," for the organization. Those ends may need further sharpening and refinement in Phase 2.

Knowing the "ends" is only the first step. In Phase 2, the retreat proceeds to its problem-solving tasks by parsing the objectives of the retreat and selecting sub-groups, if needed, before the group begins plenary discussions and moving the retreat to an agreed-upon resolution. The "writing group" for the planning document should be identified at this stage. The management sub-group must decide on the ways and means of considering the input of the retreat participants. At this point, the group needs to consider that there are many possible paths, from present conditions to future goals. Those paths may differ in total integrated costs, in time to reach the "ends," and in the sum of quantifiable results along the way, as illustrated in Figure 5.1.

Implementing the strategic plan is a very high-value change process for the enterprise that needs to be carefully monitored and managed. Once it is embarked on, all stakeholders are likely to form opinions about how well and how rapidly

---

[1] The balanced scorecard is a management tool used in many organizations in commerce, government, and the non-profit sector to align business operations with the strategy of the enterprise and to monitor and evaluate the progress of the organization toward achieving its strategic goals.

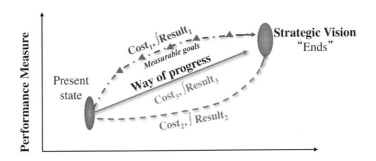

FIGURE 5.1   There are many ways from here to there.

the organization is moving toward its stated long-range objectives. To control and evaluate the change process, management must translate both mid-term and long-range objectives into readily identifiable goals and metrics. Once the change process begins, managers in the organization should be rated on meeting measurable goals, that is, in satisfying the performance metrics of the plan. They had best be able to answer the questions with respect to each intermediate goal, "How do I know when I am done?" and "How do I know if I am successful?"

To reiterate, the steps in controlling the change process of executing the strategic plan are

1. Break down the objective into specific, measurable, time-bounded components. This breakdown implies identifying a set of intermediate goals.
2. Estimate achievable quantitative progress along the course of the plan. Doing so requires specifying metrics and milestones.
3. All along the way, keep score.

An additional way of keeping score is to update the content of the balanced scorecard on an annual or semi-annual basis over the entire time horizon of the strategic plan. Besides updating the summary entries in the Table 5.1 as applied to his/her unit, a unit manager might add a semi-quantitative, self-assessment of progress or lack thereof. For example, one to three upward or one to three downward pointing arrows would roughly indicate the effectiveness of the plan implementation for the unit. Naturally, the unit manager should be prepared to defend the self-assessment to the next level of management.

Were the enterprise to have a single product and a single purpose, setting goals, milestones, and results might be straightforward. However, multi-program enterprises or those with multiple product lines have a more complex challenge in confronting the reality; namely, life in their mission space is not one-dimensional. Even single-purpose (or product) laboratories can have multiple activities (and associated metrics); the realization of which competes for resources. Examples may explain this idea more readily. Metric 1 might be the number of peer-reviewed publications about experiments (over a specified period), while Metric 2 might be the number of operating experimental stations over that same period. In a multi-program laboratory, Metric 3 might be infrastructure for biological research, while Metric 4 may be infrastructure for accelerator R&D. These activities are likely to compete for resources—both for staff and for funds—so that different resource-constrained solution sets will apply to the enterprise. The strategic plan must consider how to apportion priorities to competing priorities and their metrics.

A research organization's plans over the time horizon of the strategic plan naturally assume a certain state-of-knowledge and a state-of-technology employed to carry out its work and as a competitive back-drop for its own efforts. One might call this background environment the technological mesh that supports the enterprise's own work. Managers generally predict (or tacitly assume) that the *technology mesh* (performance level versus time of constituent technologies) will develop sufficiently rapidly to support the strategy of their units of the enterprise. Any uncertainties concerning the development of the technology mesh generates organizational risk. If the mesh is stretched too wide, it will not support the work of the organizational unit. Therefore, the competence and diligence of a manager in using techniques of technology forecasting are essential to *controlling the risks* during both the preparation and the execution of the strategic plan.

CORPORATE STRATEGIES AND POSITIONING

The *corporate strategy* defines the portfolio of activities that are most consistent with the mission of the enterprise. The appropriate mix in the portfolio for a university laboratory is likely to be far different from that of a government national laboratory, which, in turn, is generally far different from that of an industrial research laboratory with a primary mission of engineering development of products. To choose the research mix in the portfolio, the planners should consider the consistency of that mix with the strategic vision of the enterprise. Moreover, top management should direct the planners to create and then exploit readily identifiable, strong competitive advantages over other organizations. Items to consider are (1) cost of product, (2) characteristics that differentiate one's products from those of competitors (for example, quality, type of technology, etc.), and (3) market focus of the product line (in other words, niche versus broad market appeal). These concepts are not illusions. The organization and its principal investigators make just such choices every time there are opportunities to respond to calls for proposals. With respect to writing a strategic plan, the take-away is that the planners should consider the potential trajectory of the enterprise with respect to all of its significant lines of research.

The previous section counsels that a sound plan begins with gathering information and making a realistic assessment of the present state of the enterprise that traces corporate priorities back to the Vision and Mission of the organization. In Phase 2 of the planning process, the planning group begins its look forward.

For everything there is a season. In other words, what kind of corporate strategy should be pursued? And given that strategy, which programs should grow or shrink or remain as is? Schematically the progression of logic is as follows.

Vision → Priorities → Corporate strategy → Programs

Strategic category 1: The *strategy of harvest* concentrates on existing business. It seeks to maintain the acquired position in the market by retrenching and reinforcing the strongest core competencies of the organization and divesting the enterprise of other business lines while they are still profitable.

Strategic category 2: The *strategy of building* aims to develop an important domain of business activity (i.e., research) through innovation and diversification. This strategy can employ tactics of both horizontal and vertical integration by developing strategic alliances and by entering into joint ventures with other organizations.

Strategic category 3: Multi-program research institutions are likely to prefer a *mixed strategy,* in which some business lines are harvested, while others engage in building. The manner in which business lines are treated depends on management priorities based on conscious choices of institutional positioning.

Institutional positioning of the enterprise as a whole or of an organizational sub-unit shapes the priorities and selection of opportunities by addressing key *positioning questions*:

a. Who are we?
b. Who do we want to be?
c. What do others (customers, competitors,[2] and complementors[3]) think of our organization?

---

[2] An organization is your competitor if customers value your product less when they also have the other's product and if suppliers value your business less when they also have the other's business.

[3] An organization is your complementor if customers value your product more when they also have the other's product or when suppliers value your business more when they also have the other's business.

FIGURE 5.2   Areas of influence as a function of managerial position.

The potential positioning choices for R&D-driven organizations are (1) the technology leader, (2) a technology follower—always second or third to market, (3) the market leader (largest market share), (4) the deliverer of the highest quality product, (5) the most flexible organization, (6) the most responsive enterprise (when the customer says "jump," we always ask how high), (7) the lowest-price organization, and (8) the organization with the broadest offering of business lines or competencies—in other words, the place for "one stop shopping."

If you are the manager of a sub-unit of the enterprise, your vision for that unit must conform with your organization's corporate policy as it applies to your sphere of influence. Figure 5.2 indicates the correlation of management level with management tasks.

Theorem: Never march your troops into battle without an exit strategy.

CRAFTING THE PLAN DOCUMENT

Once the planning group has agreed on strategic objectives and the set of quantifiable goals and metrics of progress, Phase 2 of the strategic planning is complete. In Phase 3, writing the plan, a sub-set of the planning group begins to put the strategic plan into a written form that will be communicated to all who must execute the plan or who are key stakeholders. Writing the formal plan is a pivotal sub-project that must be done well. Top management should appoint an editor provided with adequate secretarial and graphics support to manage the writing of the document as a formal project. The final written output will be most readily acceptable to the entire planning group as well as to top and senior management, if writing is carried out in discrete steps with reviews by the entire group and by senior management after each step. Moreover, such a formalized procedure will avoid massive rewriting and its attendant waste of time and writers' egos. The flow of writing and review may look as follows:

Outline $\Rightarrow$ Review by planning group $\Rightarrow$ Storyboards $\Rightarrow$

Review by senior managers $\Rightarrow$ Key Graphics $\Rightarrow$ Review by

writing group $\Rightarrow$ Draft plan + graphics $\Rightarrow$ Review $\Rightarrow$ 2nd Draft $\Rightarrow$

External review $\Rightarrow$ Final draft.

The initial outlining and storyboarding steps should be completed rapidly, ideally each within a single week. Formal group writing is discussed further in Chapter 8.

Any written plan must have an executive summary that is easily articulated by top management to the governing body of the organization and to other key stakeholders. Following the executive summary, a clear, unequivocal statement of the strategic foundations of the enterprise informs all readers of the highest-level view of where the enterprise is headed. The third section of the plan describes the present status of the enterprise, that is, how key groups of stakeholders, both internal and external, view the plan. These general statements lead to section four: the presentation of specifics of the gap analysis, the general strategies for each principal research line, and the assessment of the technological risk and mitigation strategies over the time horizon of the plan. The action plan of section five describes programs, their objectives, and the relevant metrics by which progress during implementation of the plan will be measured. The final section describes the plans for communication and implementation of the strategic plan. It may also include details of the periodic feedback loops that keep the implementation on track.

With this document, the top management has its plan for the future. How will it manage change to get to that future?

Theorem: Plan from the top down; execute from the bottom up.

Consistent with this theorem, the executive (decision) system of Figure 1.1 crafts the strategy, enunciates the strategy, and identifies those organizational units charged with executing the plan. Those units must estimate budgets, funding sources, staffing levels, and procurements necessary to carry out the plan. They then must deploy those resources to execute the plan. The information systems of the organization have the tasks of gathering and dispersing essential information to the executives and to the managers of the operational system of the enterprise. When combined with the assessment of progress with respect to the metrics and milestones of the plan, that activity forms the control and feedback loop for the execution of the plan.

RESOURCE PLANNING FOR IMPLEMENTATION

Imagine that you are a middle manager, and you are invited to the weekly meeting of senior managers with the CEO. The CEO asks you if your unit will lead the implementation of the strategic plan. Don't say "yes" until you know the price of doing so and are given the necessary resources; instead ask for time to assess the situation and to report back to the CEO.

Your strategic questions should be: What resources are needed? To answer that question, you need estimates of money, people, services, facilities, and space. (Don't forget about space; it is in short supply in almost every organization.) You should know how resources are to be acquired. Money could come from profits, contributions (grants), gifts, or royalties. Your expanded staff might be acquired through hires, transfers, partnering, or outsourcing. You will need certain services either from internal sources for which you can negotiate or from external sources that require a procurement action. For facilities and space, you will likely have to cajole, bully, or beg.

To know how to allocate available resources, you will have to develop resource-loaded plans and schedules for work packages that you assign to your staff. *Resource-loaded schedules* summarize the estimated values of resources that are needed plus the time periods during which they are required. Although resource-loaded schedules are best prepared by experienced personnel, you need to develop your own instincts about estimates.

The basic methodology is the following. First, specify what is included and what is not included in the entire project. The most commonly used technique, the work breakdown structure (or WBS), breaks the project into clearly defined elements (sub-tasks); each one of which is owned by (and is the responsibility of) a single task manager. The lowest-level elements of the WBS are characterized by relevant parameters such as (1) work hours, type of labor (crafts, engineer, physicist), and their respective labor rates; (2) materials (unit specifications, number of units); and (3) indirect resources (procurement, legal support, communication support). For each of the elements in the WBS, the task managers then collect relevant experiential data such as previous similar tasks, vendor quotes, and so on. They must also estimate uncertainty levels. This specification is called the *basis of the estimate*.

From those data, the project manager can construct a cost model of the project (namely, implementing the strategic plan) by summing over all tasks, using scaling relationships where necessary. The uncertainties in the estimates imply the need for contingencies in schedule and cost that can be assigned using experiential data such as that in Table 4.1. These contingencies must also be summed over all tasks. The sum of estimated costs plus contingencies is the most probable cost of implementing the strategic plan.

Task managers provide the input for the master resource-loaded schedule by plotting the time dependence of each type of resource that is needed for their respective WBS element, as illustrated in Figure 5.3. This information is summed over all activities to determine the integrated time and resource deployment profile for each distinct resource type.

None of this analysis happens automatically; therefore, the project manager needs to have identified a team, including someone with good project tracking skills, to perform this analysis quickly. Typically the top management of the enterprise will want to validate the cost and schedule estimates through the use of a third party (external) review. Finally, top management needs to allocate budget authority to you (the project manager) before you can distribute it to your work package managers through their respective organizational units that have been assigned to support the implementation project. More about distribution of budget authority is discussed in Chapter 6.

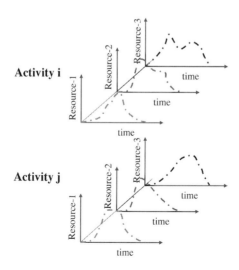

FIGURE 5.3   Scheduling several resource profiles for implementing the plan.

INFORMATION SYSTEMS

The implementation of the strategic plan is a project that serves the ongoing life of the enterprise as a whole. The collection, analysis, and communication of data are essential aspects of the corporate feedback system that allow top management to have contemporaneous pictures of the state of health of the enterprise. After all, they are managing change in the organization, not just piloting a ship. The information system of the organization provides verifiable data to assess employee morale, endurance, confidence, and trust and to record human resource incidents, near misses in worker safety, and lapses in the quality of work product. Verifiable data are key to manage expectations and perceptions of both internal and external stakeholders. Bluntly put, senior managers need a reliable intelligence network.

The world in which the enterprise is embedded is a world in flux. Externalities change. As the nineteenth-century Prussian military commander Helmuth van Moltke noted, "No plan survives contact with the enemy." Therefore, top management should continually self-assess and ask, "Is this where we really want to go?" Top and senior managers need to be aware of new technological trends and pressures, realizing that new decision criteria are likely to appear as applications of the technology evolve. Senior managers must keep their eyes open both for new opportunities and for new threats, and they must communicate their observations upward.

Despite the effort and expense of crafting a plan, *don't* be wedded to the vision out of pride. Staying the course is an option, not a dogmatic imperative. Therefore, senior and middle managers must nurture flexibility and resilience in their organizations and prepare their organization to turn on a dime if need be. Such a turn cannot be expected to succeed without the confidence and trust of one's employees.

Theorem: Keep your friends close; keep your enemies closer.

Financial Management

6

The previous chapter emphasized that crafting a strategy only establishes a picture for a possible future for the enterprise under the assumptions that sufficient tangible resources (1) exist or can be obtained, (2) can be made available to implement the strategy, and (3) can be used to conduct the operations of the organization in a manner consistent with the strategy. Item 1 requires a sound financial estimate, while items 2 and 3 require a *financial plan*. Having the plan is not sufficient; the organization also needs a highly competent resource management and tracking infrastructure that can ensure the efficient use of available resources. To put it simplistically, the financial plan is more than the CEO giving every sub-unit manager lab space and his or her own checkbook.

As with any other planning exercise, the first step in crafting the strategic financial plan is assessing the present financial status of the enterprise with brutal honesty. At every level of the organization, managers must assess the quantity and quality of their suite of resources relevant to getting their respective jobs done and must report the status of their resources at frequent intervals. Each form of tangible resource requires an expertise and appropriate tools:

Funds $\Rightarrow$ Financial officer to produce accounting reports and financial forecasts

Time $\Rightarrow$ Project manager to create work plans and measures of progress

Appropriate staff $\Rightarrow$ HR expert to generate activity reports and to schedule interviews, performance evaluations, and employee counseling

Capital equipment $\Rightarrow$ Super-technician to schedule operations and maintenance and to track work orders

Space $\Rightarrow$ Coordinator and negotiator to maintain floor plans and conduct inspections

Institutional infrastructure $\Rightarrow$ Senior management to tell you how much of it you can use.

A mid-level unit manager needs to be some of each. As a *resource manager*, s/he will employ a common measure, namely, money—the obvious measure of funds. The modern management tool that denominates time as money is the earned value measure and the schedule variance. Other measures may be related to

> Appropriate staff: Differential cost of assignments, *make–buy* differential
> Capital equipment: Differential cost of assignments, make–buy differential
> Space: Rents, utilities, and maintenance costs
> Institutional infrastructure and top management commitment: Priceless

These money-denominated quantities are used to generate estimates of the *total cost of doing business* per fiscal period (month, quarter, or year) and the *incremental costs* of changes to operations. Making the estimate must be done with considerable care as some items are easily overlooked.

The base resources of the enterprise do not appear overnight when a contract is signed. They are neither 100% assigned nor assignable; they must be maintained, and they require management. In addition, base resources generate costs that cannot be traced to *deliverables* in an economically feasible way. Such costs are known as *overhead*. One example is the cost of an on-site security force. Overhead costs must be collected in a legally defensible and consistent manner. For an enterprise to be competitive, senior managers must work to minimize overheads. As overhead functions have a tendency to expand, that task requires constant vigilance.

BASIC ACCOUNTING DEFINITIONS

Major projects and operational programs commonly require efforts across several organizational sub-units to be accomplished. Their primary budget authority is the task or project manager—or for research grants, the principal investigator. For the project to be executed in the sub-units, budget authority at designated levels must be allocated across organizational lines to business units, say Unit J, where it is further allocated among the several Activities (i), $i = 1 \ldots n$, being executed by Unit J. Thus, the total allocation to Unit J is

$$\text{Unit } J = \sum_{i=1}^{n} \text{Activity}(i).$$

For very large operating programs and product lines (uncommon in research enterprises), the executing unit may be split off as a separate corporate division.

Define a *cost-object* as any activity or item for which a separate measurement of costs is desired. Costs are assigned to objects in the following way: *Direct costs* can be traced to a given cost-object, product, department, etc. in an economically feasible (credible) way. They are those costs spent on resources used specifically for the performance of tasks associated with the project. Included in this category are (1) the salaries of all personnel assigned to work on the project and (2) the costs of materials, travel, equipment, subcontractors, etc. Estimates of direct costs are derived from the activity schedule and resource requirements for the activity. A manager can directly affect the estimates of direct costs by changing the resources to be used or the schedule for their use.

Indirect costs cannot be traced to a given cost-object in an economically feasible way. These overhead costs are *assigned* to a cost-object. Overhead expenses are costs specifically relating to the enterprise but which are difficult to subdivide and allocate directly (such as employee benefits, rent, furniture, fixtures, and equipment). They include General and Administrative expenses (G&A) and other costs incurred to keep the organization operational. Those costs include salaries of staff in the contracts and finance departments, accounting and legal services, and top management. Finally, they include profits or fees (if applicable).

The indirect costs that are assigned to specific activities are estimated by applying an established rate (percentage) to some portion of the respective direct costs. Top management and the chief financial officer (CFO) set the indirect cost rates to be applied for the entire enterprise. Managers at all levels must be aware of currently applicable guidelines.

A *cost-driver* is any factor whose change "causes" a change in the total cost of a related cost-object. For example, the cost of superconducting dipole magnets is the principal cost-driver of the cost of a hadron supercollider. *Variable costs* change directly in proportion to changes in the related cost-driver (such as the B-field in the dipole magnets). In contrast, *fixed costs* remain unchanged for a given fiscal period regardless of changes in the related cost-driver. Some costs assigned to an object or activity may have a mixed character; in other words, parts are fixed and parts are variable. Costs that vary in broad steps are sometimes called "semi-fixed costs."

To craft a time-dependent financial model for out-year projections, the main assumptions needed to define both fixed and variable costs are the cost-objects, the time span over which the model is to be valid, and the functional form of costs with time. In a model that describes how a cost-driver affects the total cost of a system, the size of the driver cannot be varied over an arbitrarily large range. Therefore, one needs to specify the relevant range over which a specific relationship between a cost and a driver holds.

TOOLS FOR FINANCIAL MANAGEMENT

To track the time dependence of cash flow and the use of non-cash resources in the enterprise, organizations use auditable financial tools that are standardized across the enterprise. Any manager having a responsibility for the expenditure of resources should become a sophisticated user of financial statements and have a sense of the limitations of the data in the statement. The fundamental tool that even first-line managers use is the *accounting system* of the enterprise. An accounting system is a formal means to gather, organize, and communicate quantitative information about the organization's activities. The accounting system should meet the demands for information from managers at all levels as well as from both internal and external auditors. External parties primarily look toward financial accounting statements to audit, while internal parties need accounting data in a form that facilitates making operational decisions. Such statements are at the heart of managerial accounting. A third form—the tax accounting statement—reports categories of gross incomes, expenditures, and changes in inventories to governmental taxing authorities at regular fiscal periods as specified by law.

As seen by the manager, financial accounting translates business events into the financial statement of the enterprise. That statement provides information about economic resources, claims on those resources, and changes in them. Such information is helpful to project the amount, timing, and uncertainties of future cash flows and resource demands. It is also useful in making investment and credit decisions to stakeholders who have a reasonable understanding of the enterprise's business and economic activities.

The sophisticated user of the accounting system understands the information flow from financial events (activities) through the applied *cost accounting standards* that embody the rules and regulations that govern the enterprise. Cost accounting standards (required of incorporated entities in the U.S.) require that the management's choices in accounting methods be applied uniformly across the enterprise. They may not be changed from year to year without full disclosure to both the governance board of the enterprise and relevant government agencies (such as taxing and securities authorities).

These options reflect the four methods of measuring income in general use: The most popular choice is the historical cost of assets denominated in nominal currency as spent. A second possibility is recording the current replacement cost of the asset in nominal currency. This option places the focus on operating income—the excess of revenue over the current costs of the assets consumed. Other possibilities are recording the discounted historical cost of the asset (i.e., in constant value currency) and the current discounted cost of asset (or its equivalent). For example, it makes no economic sense to value an antique electronic controller built in the 1950s and described on inventories as a computer that cost $25,000 to produce (in the 1950s) when $500 will buy a computer 10^8 times more capable today. Similar considerations apply to selecting one of the three or four possible ways of valuing inventory.

Even after these rules have been explained, a new line manager might still wonder why one needs managerial accounting? The answers are rather obvious. It provides information to track expenditures and progress (scorekeeping and variance analysis). It can draw attention to problem areas in operations such as overspending by subordinates. As a problem-solving tool, it can guide internal investment decisions (such as pricing and product mix) and furnish senior executives a tool for evaluating the cost-effectiveness of the operations of the enterprise and the effectiveness of internal incentives. In short, managerial accounting reports are the pieces of information to be compared against the budget plan. The reader is strongly encouraged to read an oft-cited paper by Stephen Kerr, "On the folly of rewarding A while hoping for B."

A second essential tool to manage tangible resources is the budget plan. Budgets compel managers to think ahead and to have contingencies to address for changing conditions. They aid managers in coordinating their efforts and direct their attention to problem areas that need correction. In other words, budgets articulate definite expectations that are the best framework to evaluate performance and to assure that the manager does not spend money that is not in hand.

The strategic plan defines the most forward-looking budget setting the overall goals and objectives of the enterprise. Following from the strategy is the long-range plan that forecasts financial statements for a 5- to 10-year period. Coordinated with the long-range plan are capital budgets that provide details of expenditures for facilities, equipment, and other durable investments. Finally, delimiting day-to-day operations is the operating budget.

While such budgeting plans apply quite naturally to ordinary business operations, their application to research organizations may seem more problematic. Government laboratories and nonprofit research organizations often have more difficulty implementing management control systems, partly because their outputs are more difficult to measure than are the commodity goods produced by manufacturers. Still, top and senior management need some sense about whether the enterprise yields its sponsors a good return (output) on investment (input). Therefore they invent measures such as the expenditures needed per publication in high-impact professional journals.

In research organizations, one encounters two general methods of formulating operating budgets. The traditional method takes the income for the fiscal period and divides the funds among the total cost of personnel to be supported (including indirect costs), direct operating expenses (supplies, travel, electricity, phones, etc.), and identifiable procurements. The *work package* (activity)–based budget also starts with the income for the execution period and divides those funds among activities to be performed in that period. Each activity incurs costs: (1) the cost of personnel for each sub-task (including indirect costs), (2) associated direct operating expenses, and (3) associated procurements. Summing over the work packages, the manager knows which work packages s/he can commit to and which work lacks sufficient resources to complete.

Such budgeting sounds simple at first blush, but a closer look reveals potential wrinkles. One is that the income for the fiscal period may not all arrive at the beginning of the period. Often it arrives in periodic installments (for example, quarterly). But the manager is not allowed to run in deficit[1] at any time and must only charge labor against accounts on which they are actually working. Remember that large and frequent charge transfers are a bright red flag to auditors and can also carry criminal liability.

To make matters worse, procurement of some items requires long lead-time orders. However, once the purchase order is signed, the money has a lien against it. Those funds must be considered spent even if they physically still reside in the organization's accounts. At the same time, employees expect to be paid at the routine regular intervals. The final complication is that the manager's job is not just to avoid deficits, but also rather to complete the *deliverables* on schedule.

To manage these considerations, budgeting requires careful planning well in advance. The assistance of a highly competent accounting assistant is invaluable. To plan the budget, begin with labor costs; they are the most common cause of cost overruns. First, determine what constitutes the total costs attributable to labor; those costs are not just salary, fringe benefits, and overheads. Each employee also requires materials and supplies (computers, network access, travel, "pencils," computer-aided design software, etc.). Each employee also generates distributed fixed costs (space, electricity, phones, etc.).

The cost of labor to deliver a work package is reduced if the manager can get better performance (higher productivity) from the staff. Nothing runs up costs more than not having right staff at right time. "On-the-job training" may no longer be affordable. Superior performance comes from optimizing staff assignments and adequately reviewing designs before fabrication or experimentation begins. Therefore, if practical, reassign staff for efficiency (being careful of adverse effects on morale); possibilities include sharing staff between units or activities. Finally, the manager should analyze the consequences of

---

[1] There may be very limited exceptions to this restriction if the funding agency allows the use of small amounts of bridge funds for short periods of time.

changing the resource mix; two technicians might be more cost-effective than one physicist.

Finding: When >80% of your costs are attributable to labor, you will move from crisis to crisis.

Once labor assignments and mix are optimized, consider other cost reduction techniques. Is better performance from equipment possible by regrouping activities for efficiency to minimize downtime and to maximize operator efficiency? Standardization of products (activity) where possible can reduce costs per unit; for example, design an item once and place bulk orders, etc. Consider *value and process engineering* the work packages in the program or project. In other words, analyze if equivalent functionality can be provided more cheaply or if the work process can be done differently at greater cost efficiency.

The last resort to control costs is to descope in some manner, if possible. Your grant or contract promises something (deliverables) to the sponsor. What had you promised? Everyone has heard the complaint, "We asked you for the time; you sold us a gold watch." Consider the customer's level of expectation; it is preferable to "get a B than an F." Therefore, tie all costs to deliverables, not to staff support. The work package approach ties all cost projections to definite activities that produce deliverables, making it practical to apply project management discipline to program management.

Axiom: Promise less; deliver more.

COSTS AS A DECISION-MAKING TOOL

To understand the nature of costs as a decision-making tool, two more definitions are useful. Economists often speak about o*pportunity costs*, which designate the maximum amount that would have contributed to profit had the business's limited resources not been used for an alternative purpose. If the business had used its resources for what it considered to bring the highest return on investment, then the opportunity cost is the value of the next highest value alternative.

As even a top dog cannot chase every car, the effective research manager will compare the income effects of alternatives by analyzing the associated opportunity costs. The same logic applies to staffing assignments. This consideration also applies to the not-for-profit world for research enterprises when it considers which expert to hire, which piece of equipment to upgrade, and how to spend funds designated for general plant improvement. The opportunity cost is not dispositive, but it should be a factor in the manager's decision.

The second important concept that any manager must fully understand is that of the *sunk cost*, that is, a cost already incurred. As the resources used for the *sunk cost* have already been expended, the size of the sunk costs is irrelevant to the decision-making process going forward. Nevertheless, the public often hears the contrary from politicians who continue to insist on expending lives and treasure on hopeless military adventures so that "our fallen heroes will not have died in vain."

The concept of sunk cost is critical to understanding the logic of *make–buy* decisions. In computing the *differential cost* of the alternatives to make a needed item internally or to outsource its production, the size of the sunk cost is irrelevant. What is important is the cost to be incurred going forward. Other considerations can and should also influence a manager's final make–buy decision: the quality of the product, the time to receive the product, the know-how of one's key staff, risk management, and environmental protection.

A MANAGER'S RESPONSIBILITY

Research mangers and those in the enterprise with the status of principal investigator are responsible to be aware of the cycles of research funding in their respective regions. Typically, programs take more than 1–2 years to work their way from initiatives—for innovation thrusts and for flagship programs from the administrating bureaucracy—through the legislature which must authorize and appropriate funds and back to the agencies that issue calls for proposals and grants. Managers must also be aware of the procurement rules that govern the conduct of proposers for contracts and grants (in the U.S. those are the Federal Procurement Regulations). Those requirements seem to make the rules for the submission of proposals ever more rigid in a way that can put small research groups at a competitive disadvantage with respect to national laboratories that operate major research infrastructures.

With respect to competitions for internal research funds, many enterprises issue calls to their employees on an annual or semi-annual basis. However, their award and the start of large strategic thrusts as decided on by top management must usually wait for the start of a new fiscal year to receive significant funding. To fund the implementation of a new strategic plan for the enterprise, top management should have set aside a significant pool of resources and should time the strategic planning activity so that enterprise can embark on the implementation of the plan shortly after the new plan is announced. The negative effects on organizational morale are easy to imagine if the enterprise must wait several months before inception of implementation of the plan.

Every manager is responsible for (1) good financial management practices and (2) compliance with the organization's contractual requirements. Failure to meet contractual compliance requirements can result in financial and/or criminal liabilities for the individuals involved, for the organization itself, and (in the case of contract operated U.S. government laboratories) for the Operations and Maintenance (O&M) contractor. Therefore, managers may not overspend their account and/or run a deficit. In general, they may not spend other funds in anticipation of a contract and then back charge. Money that a manager does spend must be spent for the purpose for which it was authorized; therefore, charges must be made only against appropriate accounts. Auditors see charge transfers as a "red flag."

When can spending begin? Funds may not be committed or spent unless authorized under contract with the customer, funding agency, or organization's financial officer who controls internal funds. If the organization places a purchase order or signs a contract with a vendor, that procurement is a financial commitment, which places a lien against funds in hand. Therefore, a manager must have sufficient funds to cover costs of operations plus commitments.

What does your signature mean on a purchase order or contract? Your signature is legally binding and represents your certification that you may legally bind your organization contractually. Most employees of an organization may not sign contracts unless expressly authorized by the CEO and/or CFO. Even if one only signs the cover sheet that goes to a person with designated signature authority, that signature certifies that the expense or activity is allowable and that it represents the official business of the enterprise. In addition, a manager's signature on a contract asserts that the agreement complies with all applicable laws and regulations, that relevant information and documentation is complete and accurate, and that funds are available to finance the contract.

Many customers (including most if not all governments) consider some costs to be unallowable. For example, the United States acquisition regulations explicitly identify the following items as having unallowable costs: alcohol, contributions, entertainment, gifts, interest, lobbying, and souvenirs. In many circumstances, insurance is also unallowable. The manager should also assure that all charged activities are consistent with the terms of the contract with the customer and are within the dollar limits specified in the contract. Even so, customers may dispute whether a given cost is reasonable or excessive. Therefore, managers must be prepared to defend expenditures of time and materials that they make.

Most large organizations that accept government contracts or grants are required by their funding agencies to follow detailed *cost accounting standards*. For government laboratories, these accounting rules fundamentally require that (1) indirect costs are distributed on a causal/beneficial basis, (2) all cost attribution and accounting practices are disclosed in writing, (3) the disclosed practices are, in fact, consistently followed, and (4) *full-cost recovery* is imposed in a specifically disclosed sense. Typically full-cost recovery means that a benefiting program, project, or activity is charged all direct costs and a "fair share" of indirect costs. Costs within the work scope of a customer-funded project must be charged to that project. Corporate funds must pay for all unallowable costs.

In the U.S., full cost recovery for the use of government-funded major research infrastructures (or facilities) only refers to recovering present operating costs. It does not include depreciation of the infrastructures or instruments or the amortization of the capital cost of the infrastructure. As some depreciation of the value of the infrastructure does occur over time, the funding agency responsible for the infrastructure must periodically make expenditures for capital improvement. Therefore, use of the facilities for non-public research or for foundations that do not support capital improvements can be said to incur infrastructure debt.

Sometimes for work on large infrastructure projects conducted by multiple organizations, the relevant government agency will allow (or insist on) lower than customary overhead rates even at those organizations that will not share in operating the research infrastructure in the future. Such a practice invariably incurs infrastructure debt in the organizations that are not the home of the major research infrastructure. Hence, before agreeing to partner in large projects at another home laboratory, top management should evaluate the nature of the infrastructure debt and ask if participation in the project provides sufficient residual value to the enterprise.

Theorem: Every manager has the responsibility to practice good financial stewardship, that is, to manage financial resources wisely and execute responsibilities effectively with integrity and highly ethical conduct.

The Business Plan

7

Once the business strategy is fully developed and detailed financial projections are available, the entrepreneur (or mid-level manager) is ready to ask investors (or the top management of the enterprise) for funds to launch the new business endeavor. The rationales for this should (must) be put into a written document that describes a vision for the future based upon a rigorous situational analysis followed by a strategic overview of the proposed business organization (or unit of the existing enterprise). With those foundations, a business model of the unit moves the mission, vision, and strategy into tactical action.

At universities, all have seen the simplest model of a research business within a larger research enterprise. In allegorical form, the model might be described as follows:

> A junior faculty scientist gets *principal investigator* (PI) status. The scientist hunts for rabbits that run by his cave. Graduate students gather nuts and berries. The post-doc cooks. The group eats that food tonight and survives until a new day. As the business matures, the scientist gains sophistication. The PI beats bushes close to the cave to rouse the rabbits; now the post-doc shoots arrows at rabbits that run by the cave. The PI brings back dead rabbits, and graduate students gather nuts and berries. The post-doc cooks, and the group sings the victory song and survives until a new day.
>
> The enterprise continues to grow. Now the PI and the post-doc wearing rabbit pelt clothing go off to hunt for deer. The second post-doc beats bushes close to cave to rouse rabbits so that older graduate students can shoot at rabbits that run by cave. The second post-doc brings back dead rabbits. The newest graduate students gather nuts and berries, and the second post-doc cooks. Great food! It's a great learning experience. The PI and the senior post-doc return with a small deer. Students plant a victory garden while the PI writes about hunting. The PI gets tenure.

Many small businesses do not move beyond this stage and seem happy. But growing beyond this model requires planning and analysis, either instinctive or explicit. That planning will generate the group's business plan.

Business plans are *not* just for the commercial sector, *not* just for startups, or *not* just for corporate MBAs or for top management. If you want to move your team (group, department, or division) beyond the hunter–gatherer science model or into a new business area that requires significant resources, you need a business plan. In fact, even at major research universities, a department conducting a faculty search for a junior faculty position asks applicants to submit a written research plan that will justify granting tenure several years later.

In a scientific or engineering research organization, what does one mean by a new business area? It could mean adding a large experimental component to a theory program. It might include expanding conventional research lines into "advanced" technical approaches: for example, a conventional accelerator development program might add a laser-plasma accelerator effort. It could be applying the group's basic science to a national need, for example, by transforming basic biology and materials science research into developing a novel, renewable energy supply. It might apply the nuclear physics technology of detectors to the medical/health sector by developing novel imaging systems.

The significant resources needed to make such transitions could come from either internal or external sources. A significant internal source could refer to a large fraction of the enterprise's discretionary (internal research) funds. It would likely include most of the unit manager's discretionary funds or "free energy." It will consume most of the time of the group's best staff, including the manager. External resources could come from foundation grants, exploratory governmental funding, investment by private sector partners, angel investors, or from venture capitalists (as allowed). In what follows, all these sources will be called "investors."

As mentioned at the outset, a plan that the investors will buy into begins with a vision for the future of the new business unit (or the restructured existing unit) that is based on the mission of the enterprise and is consistent with its core values. The *business opportunity* is supported by a situational analysis (SWOT) consisting of both internal and external assessments. The strategic overview of the proposed business opportunity should describe the organizational characteristics of the business unit (or start-up) and the *strategic positioning* of that organization. The organization structure will eventually have to identify the *core leadership* of the new business.

The development of the business strategy is aimed at identifying the business opportunities: What is the customer set (target market) and what benefit does our customer set need? The business proposes to make an opportunity selection for that market. In other words, it describes the product or service of the business and explains how it is different and better in some respect than the product of competitors. The bottom line of the business opportunity is the clearly articulated and justified statement of financial objectives. A compelling business plan proceeds to the tactics of implementing the strategy including the description of how performance will be assessed and how mid-course corrections will be made.

BUSINESS OPPORTUNITY

At this point, the core leadership of the nescient business needs to step back to perform a rigorous evaluation of the value proposition that the new business intends to make to potential customers and to skeptical investors. Will customers and investors find the product (research) line of the business exciting or will they see it as a boringly obvious, me too, substitution of existing products? Does the product correct serious deficiencies or defects in existing research lines from other organizations? What are the features and benefits of the product that differentiate it (in the eyes of the customer) from the products of others? Call those items *feature–benefit discriminators*. The feature–benefit discriminators must be important to the customer. The discriminators must be specific. If not, they will be ignored as mere puffery. A benefit is generally expressed as faster, better, or cheaper. Finally, a feature–benefit discriminator that is not proved will not gain your team an advantage over competitors.

Generally the core leadership needs to identify its business opportunities without revealing its interest to potential competitors—especially those with intrinsically greater resources such as research infrastructure at hand. If the business team already has the acknowledged leaders in the proposed area of research, what the group already knows or intuits may be sufficient. Trusted consultants, potential teaming partners, or current customers can add valuable information once *non-disclosure agreements* are signed. Scientists and engineers are frequently advised that strong *market pull* is more effective at gaining investors than a *technology push*. The group could assess market pull by scanning the market and by getting a consulting firm to perform market research. It should certainly remain highly cognizant of change in the research areas relevant to its product line. Finally, the team could move its product offering up the *value chain* to a level at which competition is much weaker.

The value chain is a model of business economics that describes the sequence of steps in which a business may add value to a line of products. For a research organization, the lowest level of the chain would be basic research—corresponding to the raw materials of a manufacturer. Moving up the chain would imply moving to applied research and then to engineering development and from there to full implementation in the form of a product. The top of the chain might be becoming the corporate home of a major research infrastructure.

How does the core leadership screen and filter which business opportunities to pursue? Step one is for them to align the plan with the strategic objectives of their group or enterprise (or start-up, if applicable). Step two is to assess the potential value of their product (their deliverables to the market). If the valuation is not worth the risk to the core group or to the investors, it is time to step back and reconsider whether the business concept is yet ripe for action or whether its time may have passed. If the evaluation passes through that loop, it is time to assess the constraints on the business concept—for example regulatory impediments—and threats (obstacles) posed by possible competitors. For example, competitors may have the most important customers already tied up in long-term contracts.

The process of assessment needs to be performed systematically if it is to be credible to potential investors. The assessment should answer all of the following questions, providing evidence where appropriate: (1) Who are the five nearest, direct competitors? Who are indirect competitors? (2) Are their businesses steady? Or increasing? Or declining? (3) What has been learned from their operations and publications (if available)? (4) What are their strengths and weaknesses? (5) And critically, how does their product differ from yours?

At that point, hopefully, your strong competitive advantage will be obvious. If it takes explanation, however, the potential payoff and the probability of success must justify the risk to the investors (including yourselves). If the decision of the core team is still "go," then they should map the plan onto the competencies of the staff on hand or potentially hirable in the short term. If the team is part of a larger enterprise, it should assess the balance of the new research (product) line with the remainder of the unit's portfolio of activities. In other words, it must balance the resources for the new activity against those needed to sustain the existing business. The flow of the stepwise screening (funneling) evaluation is as follows (7.1):

$$\text{Opportunity within Strategic Objectives} \Rightarrow \text{Potential Value} \Rightarrow$$
$$\text{Barriers / Constraints} \Rightarrow \text{Competencies} \Rightarrow \text{Product Line} \tag{7.1}$$

Assuming that the business concept passes through all the steps of the process, the core team must determine whether they have been viewing their idea with "lover's eyes." It is time to subject the concept to an in-depth re-examination with the help of a so-called "red team." The red team is an external group that will scrutinize the business concept by taking an adversarial or highly skeptical point of view. The red team will demand strong evidence for all the claims made by the core team.

To summarize, the business model of your group needs to describe the market, the relevant value chain, the value network of competitors and complementors, the competitive advantage of your team, an analysis of cost and profit, and the competitive strategy.

YOU HAVE A GREAT IDEA, BUT HOW DO YOU PLAN TO MAKE MONEY?

The business model of the venture is at the center of interplay between technological inputs and economic outputs. The principal purpose of the business plan is to make a clear, concise, and compelling case to investors how the new business venture will transform its scientific ideas into economic value. The written business plan describes the manner in which the business model is pivotal in a dynamic balance.

$$\text{Technological Inputs} \Leftrightarrow \text{Business Model} \Leftrightarrow \text{Economic Outputs} \qquad (7.2)$$

The business plan describes how the business concept creates value for a market by identifying its benefit/utility to the customer—the so-called *value proposition*. The plan identifies the target customers and shows how the product creates value for them. Value attributes belong to products, not to technologies; the business concept should not appear to be another "hammer looking for a nail." Rather it solves a distinct need in the *target market*.

$$\text{Technology} \Rightarrow \text{Product / Solution} \Leftarrow \text{Market} \qquad (7.3)$$

In the context of (7.3), it must identify the venture's compelling competitive advantage. In the written plan, the text should read: no one else can deliver it faster, better, or cheaper. In so doing, it should not neglect politics (especially with respect to funding from a government), and it will take account of brand loyalty such as customers preferring MIT technology to Harvard technology.

It is common for research organizations to push the development of technology as a key product. In that case, the business plan must identify the multiple ways in which the technology generates products. It will identify the most willing customers and especially those who are the most able to pay. A forward-looking plan will specify and estimate the size of the market segment in which the technology will yield greatest benefit. The place of the product in the value chain and the growth potential are indicated by identifying—with evidence—applications that are the most practical and that lead to further development. Unfortunately a great product at the wrong time is the wrong product. Therefore the business plan must explain convincingly why the present is the right time. Relationship (7.2) says that a great product must be the convergence of opportunity and solution.

Yet another question remains to be answered by the business plan. How can the venture wrap a protective layer around its technology, its product line, and its customers? A few possibilities are widely used:

1. Intellectual property: Control the knowledge underlying an innovation via patents, copyright, and trademark.
2. Secrecy: Make sure that no one else knows how to make this innovation.
3. Speed: Move rapidly beyond current competition and keep ahead. In the race, never look back; doing so costs half a step (or more).
4. Lock-in customers: Make it costly for customers to switch (or make your product the market standard).
5. Develop deep brand loyalty.

Example: Nikon and Canon provide excellent examples of using these methods. One files many, many patents; the other relies on secrecy. Both know that once photographers—especially amateurs—invest in one of their expensive lenses, they are unlikely to switch to the other brand of camera. Deep brand loyalty and lock-in is a big reason why people buy a Leica despite its high price.

The final step before writing the actual business plan is for each member of the core group to write independently the answers to the following questions. If their write-ups can come to convergence, it is time to proceed.

1. Does your idea create economic value? What is the value proposition of the product? What is the market segment? "*Market A* will value my product at level x because…"
2. Can you capture this economic value?
3. Can you protect your competitive advantage? Where in the value chain is your team focused? How will it deliver the value?

If the core leadership can answer "yes" to these questions, then it should proceed to the writing stage. Writing is an acid test of the team's thinking.

WRITING THE BUSINESS PLAN

Writing a plan demands a thorough understanding of your business concept. As the lead manager, you must own (author) the plan and so must the entire team. Operationally, the writing process instills realism in the team, documents financial (investment) needs, and lays out the operating plan of the venture. Externally, the written plan aims to attract investors, strategic partners, and key people to join the venture. As a selling document, the written plan must look and be professional but not be slick in projecting the character and the excitement of the enterprise. The text avoids anodyne truisms, boasting, and self-congratulations by the core team. Leave granting such accolades to the readers.

The principal function of the written plan is to summarize the business model and to specify the business goals of the venture. In drawing a roadmap of the venture toward those goals, the plan must identify the resources needed to stay on course as well as the timeline for their acquisition. Investors will expect their funds to be used effectively. Therefore, a well-crafted document offers a decisive operating plan for managing those resources and explains how the core team and the investors will measure the progress of the venture along the way through the use of specific, quantitative metrics and a corresponding schedule against which progress can be judged.

Unless the cover of the plan looks professional with no typos or obvious errors, potential readers may proceed no further. The cover must contain (1) a title, (2) the name of the organization (venture) and its address, (3) contact coordinates of a key person(s) or the organization as a whole, (4) in the U.S. a securities disclaimer[1] (essential), and (5) a confidentiality statement.[2] The plan is a controlled document; there is only one official version and a strictly limited number of copies. A document control number and copy number are highly recommended.

The written plan begins with an Executive Summary that is a *synopsis of the plan* targeted to executive readers. The Executive Summary is not an introduction, nor is it a preface, nor a collection of inspirational statements and exhortations. It must be Clear (logical), Concise (2 or 3 pages), and Compelling (exciting). To convince the investor to read on, the idea embodied must be too good to ignore and the financials too promising to turn down. A passionate but sane Executive Summary should enable the reader to give an attention grabbing 30-second elevator speech.

To serve all of these functions, the Executive Summary must explain the character of the organization and introduce its key personnel, encapsulating their achievements relevant to the success of the venture. The key personnel must be the credible backbone of the venture with no gaps in critical competencies.

---

[1] "This document is not a prospectus as defined by the United States Securities and Exchange Commission and is provided for informational purposes only."

[2] This is a copy-controlled document containing confidential business sensitive information. The unauthorized review, reproduction, dissemination, or other use of, or taking of any action in reliance upon, this information by persons or entities other than the intended recipient is prohibited.

As the Summary encapsulates the comprehensive business concept—vision, business strategy, and opportunity—the action plan of the venture must be explicit. It identifies and justifies the target market (or market sector) for the product along with projecting its growth potential. The bottom line of the Executive Summary is "the ask:" How much money is needed and how will it be used. The "ask" must be justified by identifying the venture's strong and *sustainable competitive advantage* and the venture's profitability or *harvest potential*. These characteristics are summarized in order of presentation in Table 7.1.

The body of the plan expands on and supports the Executive Summary in no more than 25–30 pages. More concise is better; don't increase the page count with extraneous words.

A one-page table of contents is self-explanatory. Presentation of the business concept should be built on a thorough situation analysis describing the opportunity, the new company (or venture), and the product. A claim of market demand for the product requires proof either by market research or by detailed analysis of market data that justifies the claimed economics of the business.

Without a description of focused business operations (design, development, productions, and dissemination of research results), the business case has not become a plan. Any plan built around key personnel who lack critical competencies is destined for a rocky road at best and more probably for failure. The performance metrics and schedules in the plan describe the feedback and control measures that are built into the new business. The leadership team should assume that investors would cut their losses quickly if the ventures wander far off track.

A careful risk assessment is essential in any business plan. That section will reveal whether the key personnel are realistic or engaged in wishful thinking. It should describe the likely reaction of other players in the market and especially of competitors (will they try to squeeze you out?). It considers how the venture team may react if critical external factors (new technology, materials, supplier problems) change. It also recognizes that critical internal factors can change, for example, the loss of a critical competency. As both internal and external factors can drive the venture away from meeting its milestones

TABLE 7.1 Layout of the Executive Summary

Description of the business concept and the business

The opportunity and strategy

The target market and projections

The competitive advantages

The economics, profitability, and harvest potential

The team

or schedule, the business plan must offer potential contingency actions that mitigate these risks.

The plan concludes with assessment of the finances. Investors will want to know the "burn rate" of investments on a year-by-year (or more frequent) basis. A great price of the final product if coupled with excessive development costs will price the product out of the market.[3] Investors will want to know when the first financial returns (sales or new grants) are likely to begin and when the enterprise is finally self-sustaining and growing (and if in the private sector, "in the black" and making a profit).

The entire plan is most efficiently crafted in *storyboard* form preceding the composition of any elegant prose. With a storyboard in hand, the entire team can assess the plan holistically and critically. Are there gaps in the logic? Are claims presented without evidence? Are the financials of the plan compelling? Only rarely will investors support mere charisma or a loss leader.

WHY DO PLANS FAIL?

The team should keep in mind how investors will read and analyze the text. Their first reading will be similar to the screening of resumes to narrow the field of candidates for further consideration. The venture must "make the cut" to get an opportunity to present its case in detail. The second reading must justify the nature and size of the investment being requested. In their third reading of your plan, the investors will be looking for commitment to an operating plan that both you and they can live with. If you don't "make the first cut," the investors never read the plan again.

Several pervasive reasons lead to plans failing the first cut. Technology that is too wild and unproved or that can't be protected or is too mundane will not be credible, and neither will a technology requiring an investment infrastructure that is too large for the promise. A product that is all technology push with no market pull or with an insufficient market will be immediately dismissed as not timely. Investors will demand a venture team that can articulate and execute an effective action plan. Therefore, plans that are too optimistic or too naive about potential markets or to the contrary are insufficiently ambitious will also be deemed as unworthy for further consideration. Naive assumptions and projections and gaps in analysis filled only with hand waving reflect poorly on the competency and credibility of the leadership team.

---

[3] For example, excessive development costs killed an otherwise attractive U.S. program in inertial fusion using intense heavy ion beams.

Finally, appearances are important! Even when the logic of the plan is sound and the projections are realistic, investors may be dissuaded by a plan document with terrible writing that is difficult to understand. Another path to failure is a document that is sloppy with misspellings, poor grammar, and excessive jargon or acronyms, or is just too damned long.

Theorem: Write, rewrite, and rewrite again.

Management Communication Skills

8

TECHNICAL WRITING

The written strategic plan and the business plan are two examples of formal technical writing. Their primary purpose is to communicate data, its analysis, a conclusion, and a plan for future action. In this way, they share many attributes in common with scientific writing, that is, the reporting of research results or the grant application or proposal for research funding.

> **Axiom:** Technical writing is neither poetry nor an essay nor a short story nor another form of "creative writing."

As with any writing with a purpose, the author needs to know that purpose before starting to write. For example, in the case of responding to a Request for Proposals (RFP) or call for tenders, the writer must first identify any and all pre-existing questions to be answered. In the case of a progress report or a final report for a contract, the author must realize that the writing is a deliverable to the sponsor. More generally, author should keep in mind the audience(s) for the writing. Are they the general public, reviewers, colleagues, program managers in a funding agency, a tenure review committee, or all of the above? For each target audience, the authors should consider what the audience needs or wants to get from their writing?

A written document is "for the record" evidence of the author's commitment to quality. If the writing in a scientific report is poor, the reader may infer that the underlying research is unlikely to be much better. The utility and the impact of the work are lowered, and the report will serve as a poor reference for the author, co-authors, and team members. Similarly, poor proposal writing lowers the credibility of one's team, presents a less convincing case, and significantly lowers the probability of getting funded. Consequently, any research organization needs talented editors. Moreover, the key personnel who review the reports of their subordinates should learn to write well themselves.

Whether writing for a technical journal or in response to a call for proposals, the author should use the standard format for the journal or as mandated by the funding agency. In the latter case, if the submitted response does not follow the agency instructions, the proposal is likely to be ruled *non-compliant* on its face and will not be read further or reviewed.

In both cases, read the "instructions to authors." Use good English syntax and learn to be clear and concise. (Responses to calls for proposals typically have strict page limits). Avoid "the usual sins" such as jargon, acronyms (do a final global replace with the words), clichés, and vague phrases. Cite the work of others as appropriate; avoid insufficient citation. Captions should explain the figure rather than state the obvious. A picture of a dog captioned with "this is a dog" is effectively a picture without a caption. Above all, do not tell the reader what to think of your work. Few things will annoy a reviewer more; that judgment is theirs to make, not yours.

Most technical writing contains an introductory section. First write this section, and then rewrite it after the document is complete. The aim of the introduction is to show that there is a problem to be solved. Concrete examples and citation of prior attempts are useful in this regard, but keep such background short. Then, briefly state the principal claims of the document. In a research paper, briefly indicate the originality, innovation, and utility of the approach while beginning to focus on what you are adding to the field. Do so quickly and forcefully; cut to the chase.

The following sections of the document contain relevant technical (or theoretical) background followed by a description of the methods (and instruments) used in the analysis (or research). With the stage now set, present the results of the work being reported. Surprisingly many technical reports omit a conclusion and end with a synopsis of the results. That is empty writing. What is needed at the end is a discussion of possible implications of the work (for example, results that conflict with previous research or studies) and a conclusion of what the results mean, why they are important, and what follow-on action can or needs to be taken.

GENERAL COMMENTS ABOUT STYLE

The most general advice is to use the active voice whenever possible. The bureaucratic passive voice is an effort to avoid identifying who did what. Avoid "there is" and "there are"; instead let the subject of the sentence do the work. For example, "The trace element data show ..." is preferable to "from the trace element data follows an interpretation ..." Prefer strong action verbs and precise nouns to long strings of adjectives and adverbs; in fact, try to avoid using adverbs, especially when a stronger action verb is available. Using a thesaurus is helpful.

Make the manuscript easy to read by keeping sentences short and direct. Linking sentences and paragraphs maintains the continuity of one's ideas. Stay "on point"; a stream of consciousness is neither a good technical writing nor does it make for a good progress report. Avoid platitudes, common knowledge, and useless citations; for example, don't add a citation for Newton's second law of motion unless your document is a historical treatise concerning Newton's work.

Finally and always, improve every manuscript with thorough editing. Make several revisions, editing on paper or using "track changes" in a word processing application. As you edit, ask yourself, "Can I use only half the words?" In the first draft, look for at least one correction or improvement *for each paragraph*. If you need to reread something to understand it, your readers will have even more difficulty than you. Rewrite!

After you complete the first revision, revise again. Ideally a colleague unfamiliar with the original work is more likely than you to find unclear explanations or gaps in the logic. Finally, a good copy editor will surely find typos and incorrect word choices ("there" for "their") that the spell checker overlooks. Consider this nineteen-word example from institutional plan of a U.S. national laboratory as submitted to the Department of Energy: "One important think [sic] is to avoid to destroy preserve fauna, and in particular the trees present inside the Laboratory." What the writer meant was "Preserve the Laboratory's trees" (four words).

Practical exercise: Select a paragraph of 150–200 words in the online strategic plan of a government agency. Try to reduce the word count by a factor of two or more while retaining the meaning.

Theorem: Nothing worth reading is ever written—it is rewritten.

Copyright is created directly by the act of creation of written, visual, or auditory work; it does not require any filing[1] with a governmental unit. That is in contrast to a trademark, which is issued by a governmental unit. Rules of *copyright infringement* and the legalities of use of materials originated

---

[1] Filing for copyright protection with the appropriate government agency (in the United States, the Library of Congress) does make prosecution of a claim of copyright infringement much more likely to succeed. Unlike application for a patent, registering a copyright is inexpensive.

by others vary from country to country. Infringement is any use without *express permission* of the copyright holder; that includes printing, posting on the Internet, and using the material in a derivative work. In cases of direct infringement, monetary profit is not an issue; likewise distributing someone else's work for free is not a mitigating factor.

The U.S. legal code (17 USC § 107) describes America's *fair use doctrine*. This doctrine does not definitively list exceptions to copyright infringement, but it does list those cases in which "fair use" is considered an equitable defense to infringement. In contrast with U.S. law, the European Union does not have a fair use doctrine. Instead the E.U. has issued a number of relevant directives aimed at harmonizing copyright law among its member states. The interested reader may find a list of protected rights (and citations) on Wikipedia. Consequently, one should always seek permission[2] for any use of copyrighted materials *unless* the material is in the public domain or is open access content published under a Creative Commons (CC-BY) License. In any and all circumstances, the source of the material should be cited.

MAJOR FORMAL WRITING PROJECTS—THE CDR

Most research organizations have occasions to produce major, formal documents that will require written input from multiple staff members. A structured writing process is the most efficient way to produce such documents. An excellent example of such a document is the *Conceptual Design Report* (CDR), described in this section. The suggested method also applies to major proposals or to writing a book like this one.

In the U.S. and Europe, the CDR is an essential part of the work cycle of major projects. The purpose of the CDR is to provide an overall technical description for public agencies; it includes rough cost estimates and requirements for licensing and permits. The CDR establishes a preliminary baseline description of technical sub-systems and begins the process of bringing the requirements of technical sub-systems under *configuration control*. The CDR serves as a guide to engineering for the project by providing references to technical optimization studies, links to the *work breakdown structure* (WBS), and links to the database of project-related documents. It provides background to users and stakeholders of the existing facilities of the proposing

---

[2] www.elsevier.com/about/policies/copyright/permissions.

organization and the infrastructures proposed for construction. Writing a CDR is a substantial effort; continual motivation of one's staff is generally necessary. A useful message to the team is *"Either* you design the house you will live in, *or* someone else will design your prison."

As is the case for any formal document, the writing team must understand the scope required in a CDR. This document becomes the default, *ad hoc* baseline for the machine configuration pending the initial (10%) engineering efforts that define the content of the technical design report. As such, the CDR should report a consistent overall design of the proposed infrastructure project, but it need not duplicate long analysis available elsewhere (for example, available in the organization's internal report library or in published journal articles). References to technical documentation can be included in appendices. The CDR serves as a guide to design issues and engineering challenges, describing approaches and their associated risk reduction strategies. With respect to the completion of the project, the CDR should contain a description of the baseline performance and initial acceptance criteria.

The structure may parallel CDRs of other successful projects; members of third-party reviewer panels may be familiar with several of these. The general order of chapters is (1) an Executive Summary; (2) the science case or mission need; (3) the technical chapters, the order of which should have an evident logic (e.g., start-to-end); (4) supporting research and development; (5) conventional facilities; and (6) environmental and safety considerations. The WBS of the project belongs in an appendix. Each major section should begin with a one-page technical synopsis. The format for text, figures, and tables as well as the editorial process should be set initially before any writing begins and should be consistent throughout the document.

> **Theorem:** Think before you write; think a lot before you write.
> **Lemma:** Then think some more.

The writing of sections is most effectively performed as a set of iterative cycles. The editor-in-chief (chosen by the management of the organization) should lay out the conceptual framework of the document before any text is written. Management should empower the editor-in-chief to designate the *principal authors* who will be responsible for the respective chapters of the report. Principal authors may write some sections, act as editors for their chapter, and write the technical synopsis of their chapter. Their central point of contact is the editor-in-chief. Typically, some set of senior managers together with the editor-in-chief form the *Editorial Board* for the CDR.

In Cycle 1, the Editorial Board selects the principal authors (one per chapter) and approves a Table of Contents written by the editor-in-chief. The principal authors, in turn, identify the sections and section authors (≤10 pages/author) of their respective chapters. As the writing cycles progress, they coordinate and edit sections and prioritize technical options to be presented and discussed. In Cycle 1, they review and return the Table of Contents, just adding section titles and their responsible authors. This process need not take more than a few working days.

Cycle 2 produces the outline of the CDR. Most of the effort belongs to the principal authors, who write the purpose and a two-sentence description of each section in their chapter. They note any critical technical issues and estimate the page count. The Editorial Board reviews their input and returns the outline to the editor-in-chief after deciding if any new analysis is needed.

The principal task in Cycle 3 is to identify figures to be produced and tables to be created. As graphics are frequently long lead-time items, they should be identified and assigned to appropriate staff early in the writing process. In this cycle, the section authors lay out their chapter sections (with title) in page count, identifying required content, figures, tables, and captions all in the form of a *storyboard* for each section. Figure 8.1 shows a sample storyboard (or storybook page). The Editorial Board reviews the submitted storyboard, after which the editor-in-chief meets with principal authors and their section authors to ask for revisions, if necessary. The product of this step is the Storybook (the collection of storyboards) plus the assignment to the graphics team to start producing figures and tables.

In Cycle 4, the writing team creates a mock-up of the full CDR. Now the section authors write rough (or bulletized) text within allotted space specified in the Storybook, after which the Editorial Board reviews the mock-up for consistency and to check that all critical issues have been addressed. After the review, the Editorial Board meets with all authors. In Cycle 4, the writers add links to WBS and to the document database; they should also correct draft figures and tables. So far, no detailed text has been produced, yet the CDR has a well-defined form with all critical issues identified and addressed. It is finally time to write the first draft.

Cycle 5 produces the first draft, with all authors writing the full text of their sections. Once the sections are ready, the Editorial Board reviews the manuscript sections for content, consistency, and style. The Board's review should proceed quickly if the mock-up has been properly reviewed. As sections are reviewed, the editor-in-chief meets with section authors to offer feedback. The principal authors can now write the Technical Synopses, after which the editor-in-chief writes the Executive Summary.

In Cycle 6, authors write the full text of the second draft responding to the criticisms of the previous cycle; the resulting product is the penultimate draft CDR to be submitted to the copy editors. Their edits yield the final draft, which the editor-in-chief submits for management review.

The entire process can be completed in 2–3 months as long as the writing process is scheduled and managed like a project. Although a team can produce

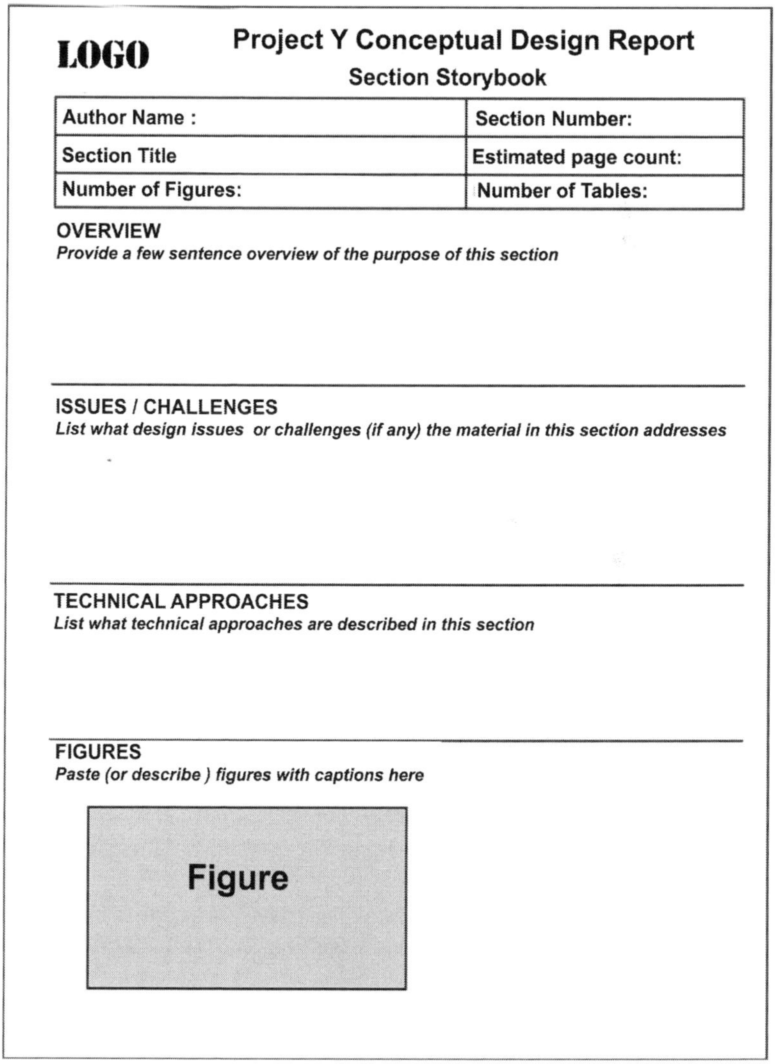

LOGO

Project Y Conceptual Design Report

Section Storybook

Author Name :	Section Number:
Section Title	Estimated page count:
Number of Figures:	Number of Tables:

OVERVIEW

Provide a few sentence overview of the purpose of this section

ISSUES / CHALLENGES

List what design issues or challenges (if any) the material in this section addresses

TECHNICAL APPROACHES

List what technical approaches are described in this section

FIGURES

Paste (or describe) figures with captions here

Figure

FIGURE 8.1 The layout of a typical storyboard for a CDR.

large formal documents in many ways, the process just described minimizes discarding and rewriting extensive blocks of text. The prize in this approach is minimizing the amount of wasted writing and ego damage to authors. The price is the major burden placed on the editors and, in particular, on the editor-in-chief.

MANAGEMENT OF MEETINGS

All managers will admit that they spend too much of their time in meetings. While some of these meeting are not under their control, some are. A manager should learn to plan and control them with as much efficacy as possible. A meeting should have a distinct purpose; this could be for an exchange of information, for planning, for making decisions, or for scheduling the operations over a specified work period. All in attendance should know what the purpose is. Meetings cannot achieve their purpose if the right people are not in attendance; additional attendees beyond that group waste people's time and can decrease the signal-to-noise in the meeting. The meeting location should be appropriate to the purpose. Some meetings are best conducted on site; some are not. For some, the local pub might be suitable. No matter where, have an agenda, set times, and specify the discussion leader for each topic. A scribe should record the action items. It is your meeting; control the timing of the topics, and balance the need to gather diverse input with the need to get the job done in the specified time.

In these times of heightened sensitivities, a manager needs to be cognizant of the cultural differences among the participants in a meeting. Of special sensitivity are perceptions of gender roles, the nature of humor, the nature of authority, and criteria/definitions of success. Quite well known are the distinctions between high-context and low-context cultures introduced by the anthropologist Edward Hall. In Hall's concept, communication in high-context cultures is less direct and depends on non-verbal communication such as body language, the nature of the setting, and emotional response. In contrast, members of low-context cultures are far more likely to express their opinions and positions directly. A contemporary example has been the position of the Japanese government regarding a decision to site the international linear collider in Japan as urged by the Japanese Physical Society. Time and time again, the cognizant government ministry has deferred a decision. Many observers take these repeated deferrals as a sign that the Japanese government will never say "no" directly: they just will never say "yes." Such behavior is frequent in Japanese culture. As scientists and engineers have frequent occasions for cross-cultural meetings, they should make themselves aware of such cultural differences.

Some meetings, especially those with a very small number of attendees (think supervisor–subordinate meetings), require that the manager show a strong, active interest in understanding others. Such behavior is often called *active listening*. The manager should look open and receptive and should try to see matter from speaker's point of view. Establishing eye contact with the speaker establishes a "listening–speaking connection." In such meetings, paying attention to the body language of others, for example, facial expressions and nodding or hand and arm gestures, can provide crucial clues to the differences between issues and motives and can distinguish between logical and emotional content of the speaker's words.

The manager should signal encouragingly to show understanding and interest in what is being said. S/he should avoid distracting behaviors such as drawing, playing with a pen, etc. A useful technique for the manager is to repeat what the speaker has said to check for the accuracy of his understanding. This *reflexive technique* is most appropriate when counseling employees; clearly it does not fit every circumstance or purpose. Meetings of this type benefit from having a third-party present (a HR representative or deputy) who can listen for the manager when s/he speaks.

Regardless of the meeting type, it should "end with a bang, not a whimper." That requires knowing when to end and knowing who (for example, the scribe) will summarize the meeting and repeat all the action items and the party responsible for each action. The manager who called the meeting should then end with a positive note and articulate who will track follow-up actions to ensure implementation.

NEGOTIATIONS (IN BRIEF)

Most meetings have some element of *negotiation*; negotiation is only a part and not the purpose of the meeting. In other words, negotiation is one form of *conflict management* between parties. Part of the purpose of some meetings is agreeing on a balance of interest among parties. An effective manager prepares for both purposes whether the meeting is one-on-one or multi-party. In what follows, a one-on-one situation is assumed for the sake of simplicity.

Any negotiations rest on some basic assumptions listed in Table 8.1.

As a representative of your organization, know the prize you are seeking and what your organization is willing to give up (the price) to get it. In addition,

TABLE 8.1 Assumptions that underlie negotiations

All parties have at least one objective in common. What are they?
Other objectives are divergent or in direct conflict. What are the areas of conflict?
Each party knows its principal objective (its *prize*).
Each party seeks some specific action from the other.
Each party perceives some benefit from talking.
Each party is willing to give up on some objectives (*the price*).

know yourself (motivations and emotional context), and try to understand the other party from his/her point of view.

Theorem: For every Prize, there is a Price.[3]

If the Prize is important to you, prepare! If not, why are you wasting your time in meeting the other party? The elements of preparation are as follows: (1) Understand issues and interests and the distinction between them; (2) understand the emotional dynamics of each of the participants; and (3) understand the power dynamics of the participants.

Regarding the Prize: Distinguish between the interests—the basic needs of each party—and the issues of the parties. Interests are often intangible and abstractions: values, reputation, esteem, sense of security. They are rarely negotiable, usually intangible, and cannot be quantified (measured). In contrast, issues are tangible items necessary to acquire, control, protect, or *to satisfy the interests* of the party. Issues are usually negotiable—why both sides are meeting—and are usually tangible and quantifiable. Interests must be satisfied to resolve the conflict between the parties.

Regarding what each is willing to give up (the Price): *Bargaining chips* are of low value to you but of great value to the other side. Your side is willing to trade these chips for something more valuable to you. However, *non-negotiable demands* are resistance points, matters that are valuable to your side but are cheap to theirs. Each side feels that it must get these if they are to reach an agreement. A schematic of the bargaining positions is illustrated in Figure 8.2.

From the figure, one can see why negotiations can break down or even worsen the conflict. Based on initial talks, one or both parties may decide that fighting is more important than agreeing. Another possibility is that the basis for negotiation crumbles because at least one party decides that there are no shared

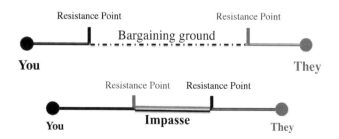

FIGURE 8.2 The geography of bargaining grounds and impasses.

---

[3] I am grateful to negotiation coach, Nicole Shapiro, for this theorem.

objectives. The world of international politics has shown instances in which talking becomes a weapon in the fight between the parties; in other words, one party decides that its objective is to keep the other talking while it acts.

If negotiating were rational, computers could do it; rather it is interplay of power, personality, and psychology. It also hinges on the degree that the parties value their relationship. Is the relationship enduring or a one-off encounter? Consequently, consultants expert in negotiations advise their clients:

1. Assess the personality types of members of the other team: What are their styles? What are their hot buttons? And who are their hold-outs ("bulldogs")?
2. Assess the power relationships of members of the other team: Who is the real leader? Whose interests are most at stake and what are those interests? What is their source of power?
3. Stay alert to dynamics during the process, both to emotions and to the flux of power.

The shadow power source in negotiations is the Best Alternative To a Negotiated Agreement. The BATNA speaks to the questions: "How well off are you if the negotiation fails to reach a compromise? How eager are you for a settlement? How eager are they for a settlement? Decide these questions *before* you start negotiations. You can get the best deals when you are prepared to walk away.

Once negotiations begin, your team should speak with one voice. Therefore, your team should agree on the zone of possible agreements and your resistance points in advance and in private. During the negotiation, guard against the paraphrase; don't let your position be moved by the paraphrase. Framing of issues is important to the flow of the negotiation; do it if you possibly can.

The topic of becoming a successful negotiator is far more complicated than can be treated in this chapter. Any manager is well advised to enroll in a detailed training course. But before ending the topic, it is useful to describe the so called *win–win success* criteria:

1. Negotiations have maximized the "value added." No other solution better satisfies all the interests examined.
2. Any redistribution of value that gives significant gains to one side will take significantly from the other.
3. All believe this is best solution under the circumstances.
4. All participants believe the process was fair.
5. All participants and their advisors agree that the settlement is legal and proper.

INVESTIGATION FOR MANAGERS

Not infrequently, a mid-level or senior manager is called on to investigate a complaint from an employee, a peer, a customer, or a supervisor. In the case of an employee's complaint/charge of misconduct, do not initiate an investigation unless there is no other option. This recommendation becomes more compelling the higher in management that one rises. The manager should avoid being the "police," the jury, and the judge; s/he must consider how to minimize organizational trauma from serious charges. Get appropriate expert assistance from HR, from the organization's general counsel, from an *Environment, Health, and Safety (EHS)* representative, or from other subject matter experts. If possible, appoint (or have HR appoint) a neutral staff member to lead the investigation. Move on multiple fronts to gather physical evidence and to question others with possible direct knowledge. Act as quickly as possible because evidence decays or is destroyed or other employees find reasons to feign ignorance.

Investigations raise communication issues on two fronts. With respect to information to other staff: (1) apply "need-to-know" principles, (2) be factual but reassuring ("We are all in this together"), and (3) keep in mind that one can destroy a lot of trust after "incidents." With respect to gathering information from staff: (1) know your objectives, (2) get the facts not hearsay, and (3) ascertain the judgment of participants and the motivation of the participants. Interviews during the information-gathering phase should stick strictly to the investigation itself; extraneous personal or social interchange should be avoided. The methods used during investigations are to ask for evidence and in some instances to interrogate.

Interrogations are best left to Human Resources or the organization's general counsel. (Doing so maximizes the legal protection of the responsible manager.) They will know and respect institutional policy regarding the presence of an employee's representative at such meetings. Unionized employees always have the right to the presence of their union representative.

THE MANAGER AS JUDGE

As an investigation draws to its close, the managers must take on the task of a third party intervening to resolve conflicting claims: the complaint and its defense. The options for handling the complaint depend on whether the complaint alleges a violation of institutional terms of employment. If it does, then a formal adjudication of the complaint allegation is in order following an impartial investigation by third party (often HR or an institutionally designated arbitrator).

Administrative action by the manager must be a fully documented continuation of a formal process that always and only follows fair process as defined in employee handbook or equivalent document. That process should include a description of mechanisms and rights of appeal. All senior managers should understand this process well, as the step beyond is legal action or external arbitration depending on the terms of the employment contract. Many institutions will provide legal assistance to their managers only if they have followed all institutional guidelines. Keep a detailed, contemporaneous record of all your conversations, meetings, and actions in such instances. Assume that all written records are discoverable in legal proceedings.

In the case of complaints within the bounds of institution guidelines and work rules, the manager's task is one of problem solving and mediation. One possibility is a direct negotiation between complainant (C) and an apparent offender (O). For example, C might send a letter or speak with O, to which O might respond verbally or in writing. Assuming that the two cannot resolve their conflict amicably, both may/should send a written notice of their position to the manager, who will act as a mediator and document the resolution of the conflict.

The manager's role as a mediator is to reduce tensions, facilitate communication, identify mutual interests, and clarify decision criteria. If the manager is ultimately to act as an arbitrator, s/he should assist by providing an integrative framework for decision-making, cutting through non-substantive issues, and if necessary deciding and documenting the outcome of the dispute. The manager's preferred role is to remain a mediator by making clear to parties that s/he prefers to facilitate their agreement, thereby increasing the commitment of parties to come to a resolution between themselves. Nonetheless, a manager should specify that s/he is ready to make a *binding decision* on a set timeline as only adjudication can authoritatively resolve some issues. That makes clear that mediation is not a sign of indecisiveness, and it keeps conflicts from festering. If the employees have come to their manager to work out difference, the manager should repay their trust by working with them to rebuild strained relationships.

As a manager, you will encounter circumstances in which you must be the adjudicator. Ask yourself if you are credible in that role.

Topic for introspection: Evaluate these questions in the context of your position:

1. Do you inspire confidence?
2. Are you seen as engaged? Helpful?

3. Are you respectful of people? Of process?
4. Are you seen as fair? Impartial?
5. Do you understand the rules? Boundary condition?
6. Are you judgmental or judicial?

Hopefully in formulating your answers, you will understand why being a successful manager requires more than subject-matter expertise.

Marketing Scientific Organizations

9

An organization's executive director of marketing (or that side of the job of other senior management) has a key role in executing the strategy of the enterprise. From the perspective of enhancing the reputation, business, and financial success of a research enterprise, the role of the marketing director (or cognizant senior manager) can be divided into two main categories: (1) developing and promoting the *positioning* of the product lines of the business and (2) enhancing and promoting the *business position* of the enterprise. For an enterprise, the primary product of which is research; these two tasks are tightly intertwined. A failed research line can deeply damage the standing of the enterprise in the eyes of sponsoring agencies and as well of the peer community—both competitors and complementors. Conversely, successful research lines are among the most important means of building the reputation of the enterprise. As the reputation of the organization is an important attractor for new personnel—in particular for "star players"—the importance of enhancing the business position of a research enterprise needs to be assigned to some senior manager[1] as a primary assignment.

MARKETING FOR SALES: IDENTIFYING A PRODUCT LINE

Marketing is frequently considered everything that is involved in the diffusion, promotion, and sale of the product. With such a broad definition, it is not surprising that engineers and sales managers often come to divergent points

---

[1] In U.S. national laboratories, that person is often the Chief Research Officer.

83

of view about marketing based on different ideas about the market. From an engineering perspective, improved features and performance are the most important aspects to develop at any price point. In contrast, a sales manager may be most concerned about wide product availability and compatibility with established price points (and their associated consumer resistance levels). Neither is wrong. Both have identified a characteristic of the product that is paramount for some subset of customers (consumers). Once one accepts that perspectives and perceptions are crucial to the company's business positioning, the means of satisfying the customers' needs follows, progressing from initial idea through market research to product concept. That logic applies whether one is considering making a consumer product or satisfying a mission need of a funding agency.

More and more frequently, funding agencies ask that designs of major research infrastructures fit their arbitrarily set *price point*; unfortunately, just as frequently, agencies demand performance characteristics that are inconsistent with their desired price point. That situation may arise because the agency (customer) is unable to articulate and prioritize its needs consistent with fiscal realities and unable to restrain its bureaucratic urge to impose project procedures that inevitably raise the project costs by 20%–50%.

In such cases, the customer may be ignorant of what will satisfy the agency's mission need as an affordable, interim product. Executives in a research organization must have the courage to discuss the price–performance trade-off with agency program managers in stark terms with firm evidence of costs and associated risks on the table. These very same considerations pertain to consumer products. Not doing so only generates false expectations that eventually lead to programmatic failures. If the agency manager—or any customer—says, "All I can afford is X; *take it or leave it*," it is time for the manager to "leave it" and walk away.

Whether targeting a market segment for research or for consumer products, the marketing (or business development) staff should have the task of identifying and distinguishing areas of the customers' actual privation (for example, a product needed to fulfill an agency mission) from items on the agency's wish list if money were no object. In short, the tasks for the business development (marketing) staff are (1) to identify and understand customers' needs; (2) to determine whether a given product can satisfy customer needs or whether a need for the product can be created, depending on the product line and potential customer and (3) to help potential customers find, buy, use, understand, and value the products of the enterprise.

If the research organization is targeting a potentially new customer, generating ideas of possible products is the essential first step. In that process, there is no substitute for creativity. A recommended procedure is to funnel possibilities: The *funneling* procedure has already been introduced in Chapter 7 and was shown schematically in Equation 7.1. Start with many ideas; test each idea and refine the product concept. Then do more sophisticated tests, and search for obstacles as you progress to a final product idea to market to the customer.

An essential ingredient in funneling product concepts is a good understanding of your customers and their needs as they perceive them. Get to know the customers well; understand their point of view about issues, including where and in what ways the performance of current products falls short. Be able to describe their priorities and their decision processes. Then, work to enhance their perceptions of your company or product. In the world of commercial products, a primary means of ascertaining customer interest is the market survey, both experimental and non-experimental.

Experimental approaches consist of testing potential products in the field or in the laboratory. The latter venue is frequently used in the food and beverage industry via tests that evaluate preference in appearance, taste, texture, shelf-life, ease of preparation, etc. Independent consumer organizations often perform such tests. In many ways, platform presentations and poster talks at technical conferences have an aspect of field-testing research product lines. Non-experimental tests include surveys, focus groups, interviews, or for research organizations observation and analysis of funding patterns, tracking broad area announcements of research opportunities, meetings with program officers in funding agencies, etc. Program development offices in research enterprises commonly employ all of these approaches. In addition to influencing potential customers, market research can be a highly effective way to build consensus across an entire research community. When that happens, even reluctant funding agencies may discover for themselves the importance of a new research area. A good market research department moves logically and explicitly from problem identification to data collection to analysis to making its recommendations to the appropriate levels of line management.

Once a product (research) line has been identified, an additional task remains that should precede any large investment by the enterprise. The relevant manager of program development in collaboration with the functional (production) department needs to determine whether the potential research line is likely to be captured immediately by competitors once the research line is made public. Alternatively, can the enterprise design a lock-in mechanism that makes it difficult for an agency to fund others to do the work? For example, one's enterprise may have a nearly unique infrastructure advantage. Can the enterprise move so fast and capture market share so fast that it is not profitable for others to compete without making excessive expenditures? Can the enterprise enroll its complementors (for example, suppliers) in a way that makes it unattractive for them to give volume discounts to competitors? An enterprise that is first to market and advances a research line quickly can be difficult to catch unless competitors gain advanced insights and make large investments.

MARKETING FOR SALES: POSITIONING PRODUCTS

Whether a product is meant for household consumers, for technology enthusiasts, or for government agencies, many alternatives may exist that promise to fulfill the same basic function. How does the customer choose between products? Previous chapters have introduced the concept of feature–benefit discriminators. *Product positioning* rests on that concept. Different customers and different market segments will value one benefit over another; the value space is multi-dimensional. Size, weight, power consumption, reliability, serviceability, compatibility with other products, potential for upgrades, and flexibility of use are all possible considerations; price point is almost always important. The marketing manager needs to work with the product manager to decide on the optimum product positioning, be it lowest cost, most available, most durable, or top performance. Choosing the positioning of a product implies targeting an identifiable market sector.

An example may illustrate. The 28-km Large Hadron Collider (LHC) at the European Organization for Nuclear Research (CERN) must be able to rapidly (several milliseconds) dispose of its full energy proton beams in the case of a magnet failure. It does so by means of deflecting segments of the beam into exit ramps. Turning on the deflectors takes ~100 ns. In order that the beam does not destroy the septa of the off ramps, the segments of the beam are separated by "abort gaps" of somewhat more than 100 ns duration in which there must be no beam particles at a level of better than one part in 10^4. The fraction of beam particles in the abort gaps must be monitored before the stored beams are accelerated to full energy and also during full energy operation. Three technologies were considered to monitor the abort gap: Beam Position Monitors (BPMs) as used throughout the LHC, a system based on Avalanche Photo Diodes (APDs), and a laser four-wave-mixing monitor proposed by the Berkeley Lab (Table 9.1).

TABLE 9.1   Characteristics of mission critical abort gap monitors

	REQUIRED	BPM	APD	LASER 4M
Timing precision	300 ps	1 ns	100 ps	3 ps
Dynamic range	1.00E−04	1.00E−03	1.00E−05	1.00E−07
Maximum cost	$1,000,000	$10,000	$500,000	$5,000,000
Sensitivity to beam energy	Insensitive	Insensitive	Sensitive	Less sensitive
Complexity	Moderate	Simple	Moderate	Complex

Each potential product had at least one attractive feature; each had at least one drawback. The Berkeley product could have performed the technical job but at high cost and after moderate development time. The high cost also bought a unique technical capability, but one that was not needed (at the time) by CERN. The BPMs available from industry were cheap, easy, and familiar to CERN technicians, but they did lack the requisite dynamic range. The APDs had a significant drawback, which CERN thought it could circumvent given some development time and extra cost. In the end, the APDs were the best-positioned products and were chosen even though the laser system had the superior characteristics in performing a mission critical function. For a mission critical instrument, the cost should not have been an issue, but the laser system was complex, and most of all it suffered from a "not invented here (at CERN)" characteristic. Moreover, as the APD approach was mission critical, CERN was not concerned when the final cost was more than the originally anticipated maximum cost in the project plan.

As perceptions of a product can outweigh product realities, what program developers claim about a product should match both product performance and the customer's target characteristics. The cost to the customer should match the price point (but keep in mind the earlier caution). Some product positions cost more. Top performance initially costs more than "just good enough" performance even though "just good enough" hardware may require eventual upgrades. Externalities also matter; the product must be consistent with present and projected regulatory structures.

Other considerations can also influence product positioning:

1. Competition: it is better to be where the competition isn't.
2. Company skills: some positionings are better matches to organizational strengths.
3. Customers: it is better to position on characteristics that are important to customers.

Whether positioning a research line (product) or the entire enterprise, managers are tempted to think tactically. For example, position toward the average customer or serve the most common customer needs. Particularly for senior executives, it is more effective to think strategically. Find the strength of the product or the enterprise in terms of benefits to the customer: own an image. If possible, avoid price competition by owning a set of customers; in effect create a "local monopoly." And always think in terms of feature–benefits that differentiate the enterprise and its products from those of competitors; benefits offered yield customers served.

STRATEGIC MARKETING

In contrast to marketing for sales, *strategic marketing* means making choices to enhance the *business position of the organization*. That concept has already been introduced in Chapter 5, which described the various positions that an enterprise can take, from technology leader to lowest-price supplier. To be clear, strategic marketing is not just advertising one's enterprise or claiming that all research done by other labs is also a core competency of one's own lab.

Exercise for discussion/introspection: Analyze in bullet form a recent or proposed "science initiative" (that is, a product idea) from your organization. Identify the need and the consumers for the product. Identify what—if any—market research was done and what marketing materials were created. Does your organization have partners? Why or why not? Why does your organization fit market needs? Does your organization have a marketing strategy for that product?

While positioning of the enterprise is an important aspect of strategic marketing, it is only one part of its primary functions: (1) finding and creating business opportunities, (2) creating competitive advantage for the enterprise, and (3) challenging the competitive advantage of others. In short, a marketing strategy for a research enterprise is an integrated set of deliberate choices about how to create, capture, and deliver value to customers (funding agencies and foundations) over long periods of time. Value is created by the willingness of customers to pay the opportunity cost of the research enterprise (their supplier).

Effective marketing strategies are based on resource advantages of the enterprise. These advantages may be created by speed or by gambles, often under the cover of competitor ignorance. Two simple, but critical, points should be part of any marketing strategy: (1) Base your strategy on the strengths that differentiate you from competitors. Those strengths are your resources. (2) To decide on a strategy, you must identify those resources. You may find that it is unnecessary to worry about creating more, or new, resources. You create value when you do what no one else can. Advantages that are difficult for competitors to secure are listed in Table 9.2.

How can an organization attack the competitive advantages of competitors? Mere imitation of a competitor's resources is unlikely to lead to a large win. Moreover, it typically requires already a large enterprise to succeed.

TABLE 9.2 Potential resource advantages that competitors can't get

COMPETITIVE ADVANTAGE	SOURCE OF ADVANTAGE
Unique and owned by you	Patents, other IP
Unique and can't be sold	Superstars, major infrastructure
Unclear how to create it	Regulatory barriers
Not clear what it is	Know-how
Enhanced returns to scale	Experience

An exception is to exploit a niche that is not well covered by the competitor, but mere imitation always remains an option. More likely to succeed is developing substitute resources or production (research) technologies with different attributes that produce a "faster, better, cheaper" product and to do so in a manner that is difficult for others to imitate.

A further means of securing advantage is forming strategic partnerships (discussed in Chapter 14) and alliances that can upset the *resource fits* of one's established competitors by synergistically enhancing the suite of resources of one's alliance partners. In that way, the alliance becomes able to challenge the competitive advantages of other businesses (or laboratories), including by expanding the set of loyal customers.

Research Ethics

10

All managers must be aware of the ethical issues and conflicts in their respective spheres of influence in the workplace, because the enterprise depends on their abilities to resolve such issues. Some ethical conflicts may arise when work duties clash with the beliefs of conscience of individual employees. Potential conflicts could arise from the prior acceptance of a policy that had been a condition of employment. In the 1950s, 1960s, and 1970s, all employees of nuclear weapons laboratories in the United States were required to state their willingness to accept assignments in a weapons design division should the Nation require their service. Not all applicants were willing to accept that condition.

Far more commonly, managers are given tasks of hiring employees (or making hiring recommendations), judging the professional competence of employees, disciplining them as recommended by Human Resources, and recommending or awarding levels of compensation or employment status (promotions). In organizations with a large population of students, enforcing obligations toward the safety and well-being of students is a paramount responsibility. Obligations to the public with respect to the environment, safety, and stewardship of publicly owned resources call for a keen sense of doing what is right and not just what is minimally mandated by law. Finally any manager in a research organization must be vigilant about recognizing and eliminating any and all scientific misconduct or the appearance thereof.

When a manager acts in any of these areas, his credibility with employees and with top management rests on perceptions of his character, temperament, and the rigor of the processes that he employs. In synoptic form, requirements in each of these areas are as follows.

Perceptions of character: The manager is fair; s/he acts with integrity. S/he is wise (or at least competent) and respects the confidentiality of all parties (consistent with the rules and regulations of the enterprise). An exception is that managers must generally report all complaints of work-rule violations (especially harassment) to HR.

Perceptions of temperament: Managers are expected to listen carefully, to withhold judgment until all facts are collected. Both supervisors and subordinates expect managers to understand and act in accordance with the applicable rules and laws. Managers should not "fly by the seat of their pants"; they should obtain expert opinion when it is needed. In the final analysis, the manager should be seen to balance all considerations.

Perceptions of fair process: Everyone expects that processes of investigation and adjudication will be fair and impartial. All managers, and especially senior managers, should strive to notice all elements of any alleged offense and to conduct or oversee investigations consistent with the relevant rules, policies, and established practices of the organization. The actions of excellent managers are timely; an old adage states, "Justice delayed is justice denied." Fairness demands that all conflicting parties have the right to present their own evidence and to respond to concerns raised by the other parties to the conflict. Typically (and as specified in the employee handbook) managers should grant employees a right to an advisor, and in some cases, legal counsel. By the legal contract between the union and the employer, any unionized employee *must* be given the right to representation by the designated representative of the union (shop steward).

The credibility of a manager as being fair to all employees depends on his/her being impartial in fact-finding in hearings, and in rendering decisions that are even-handed, not capricious, not unreasonable, and not arbitrary. When giving notice of the decision, s/he also supplies the supporting reasons. Fair process contains a mechanism of appeal, freedom from retaliation, having one's case treated the same way as similar cases, and privacy, as far as possible, for all concerned.

The text *Ethics of Emerging Technologies*: *Scientific Facts and Moral Challenges* by Budinger and Budinger recommends "the four A's" as a systematic, pragmatic approach to handling ethical conflicts:

A1) **A**cquire the facts: Get as many facts as possible, define uncertainties, clarify ambiguities, and seek advice from others.

A2) **A**lternatives: List alternate solutions and develop alternate plans in parallel.

A3) **A**ssessment: Assess possible solutions according to normative theories[1], identify and prioritize the interests of stakeholders affected by the decision; perform risk analysis when appropriate.

A4) **A**ction: Decide on a plan or plans for action; keep alternate action plans under consideration; adjust and adapt, recognizing that an initial solution may require revision; keep an open mind to new options.

---

[1] Many professional societies promulgate ethical codes, for example, engineering, medicine, and clinical psychology. These codes may provide useful guidance.

The "four A's" approach does not yield *the* answer, but it does yield a considered, defensible answer.

> *Topic for introspection or discussion*: The ethics of unintended consequences. A majority of professors with large research grants at major research universities as well as senior scientists and administrators of major scientific laboratories with large, relatively stable annual budgets have driven the overwhelming push for "Open Access" scientific journals in which the author pays for having a paper considered for publication. Without question, this model puts those with tiny research grants such as early career researchers, students, and faculty at small colleges and universities at a competitive disadvantage with respect to building an impressive publication record. Are the proponents of Open Access guilty of an unethical conflict of interest?

ETHICAL ISSUES OF SCIENTIFIC RESEARCH

An ethical issue that is specific to the research enterprise is *scientific misconduct*. This term includes a variety of offenses that may be committed in all aspects of the research process, that is, in proposing, performing, or reviewing research or in reporting research results. Some definitions are in order.

a. *Fabrication* denotes making up data or results and recording or reporting them. It also refers to deleting data without stated justification. Where photographs are primary data, one may not scale up (increase the pixel count) an image submitted for publication; interpolated pixels are not data.

b. *Falsification* includes manipulating research materials, equipment, or processes such as making significant departures from accepted practices. "Significant departure" and "accepted practices" are subject to interpretation; claims of such falsification must be substantiated. Significant departures should be documented, described, and justified by the researchers. Recognize, however, that significant departures are frequent in breakthroughs. Falsification also refers to changing or omitting data or results in a manner that the research is not accurately and completely represented in the research record. This latter category of offenses includes performing inappropriate statistical analyses or selectively altering and distorting photographic data. *Overall and uniform* brightening, changing the white-balance, or contrast of an *entire* image are generally not considered falsification *unless* those characteristics are in themselves the primary data of the image.

c. *Plagiarism* is the misrepresentation of another person's ideas, writings, images, software and hardware processes, or results, without giving appropriate credit. This definition of plagiarism does not distinguish between one who conveys ideas without attribution and one who presents information as if he were the originator. Self-plagiarism refers to representing one's work as new and original when one has reported the work previously without acknowledging that fact.

d. Violations of research regulations include performing or knowingly participating in unauthorized human or animal experiments. Conducting research that knowingly endangers public safety is likewise misconduct.

e. Failing to report wrongs when there is a responsibility to do so is misconduct in itself. Managers must document, investigate, and report any allegations of misconduct; otherwise they can be found complicit.

Honest error or differences of opinion may be cause for a manager's concern, but they do not constitute *research misconduct.*

Research misconduct is a very grave charge; a substantiated finding of misconduct is considered *just cause* for dismissal (even of tenured employees) by many organizations. Because of the gravity of the charge, the evidentiary standards for a finding of misconduct should be and are rigorous. The U.S. Office of Management and Budget, in promulgating its Federal Policy on Research Misconduct requires that misconduct (1) be a "significant departure" from "accepted practices"; (2) be committed intentionally, or knowingly, or recklessly; *and* (3) be substantiated by proof by a "preponderance of evidence."[2] Read carefully and follow the policies of your state, country, and home institution regarding research misconduct.

A gray area concerns the allocation of credit in scientific papers. Credit is explicitly acknowledged in three places: the list of authors, the acknowledgments of contributions from others, and the list of references or citations. *Citations* place the work in a scientific context; insufficient context is a frequent criticism made by reviewers. As citation of directly relevant work is part

---

[2] Preponderance of the evidence is a rather weak standard, although it is a standard in tort law and other civil litigation. It can be as weak a determination as a 51%–49% judgment.

of the reward system of science, failure to allocate *full* credit denies just reward and may lead to charges of misconduct.

Many conflicts within research groups may be avoided, if rules for assigning authorship are agreed upon beforehand. Formal agreement at the outset is a vital part of the management of large research collaborations. Any research organization, large or small, is wise to discuss guidelines for credit at regular group meetings that include students, staff, and faculty. *Authorship* implies a person having made substantial contribution to the conception and design, or data acquisition, or analysis and interpretation; just raising money is not enough. All authors are responsible for the *integrity* of the data, and each author takes *public responsibility* for the *entire* paper unless that is expressly limited in the manuscript. (The statistician might expressly not take responsibility for the pathology in a medical paper.) Never offer or accept honorary (gift) authorship; the laboratory director does not get a free ride. Technical support staff should be listed in the *acknowledgments* but not listed as authors unless they have performed all authorship duties. Finally the group should adopt a written policy on the order of authors.

The faculty–student relationship has its own special obligations. Faculty should never pressure graduate students for co-authorship credit. Notice how Gold (author pays) Open Access corrupts this concept. Graduate students should always *acknowledge* contributions of faculty advisors to their scholarly publications. Unfortunately few[3] department chairs acknowledge instances of improper authorship; faculty members think that students don't understand yet (they will understand when they are in positions of power). Meanwhile many students (39% in the same APS survey) claim that they have seen instances of improper authorship; clearly, many students feel ripped-off.

Be aware that specific criteria for authorship are discipline dependent, and these criteria are a matter of controversy. For example, the *Journal of the American Medical Association (JAMA)* asserts that authorship credit should be based *only* on (1) substantial contribution to conception and design, or data acquisition, or analysis and interpretation *and* (2) drafting the article or revising it critically for intellectual content. *JAMA* further stipulates that acquisition of funding, collection of data, or general supervision of the group alone does not justify authorship.

The U.S. National Academy of Sciences (NAS) considers these guidelines to be too strict. Yet given the lofty academic positions of Academy members, one may object that the NAS position derives from a conflict of interest. The opinions of many senior professors are closer to that of the NAS than to *JAMA*.

---

[3] Roughly 2% in a survey by the American Physical Society (APS).

CONFLICTS OF INTEREST

Engaging in professional or personal business activities that might compromise or result in less than optimal service to an employer or client raises an issue of *conflict of interest*. An area of clear conflict is an employee having a personal financial stake in business decisions that s/he makes for his/her employer. Getting paid twice for doing the same work—double compensation—is a clear conflict as is competing with your employer or consulting for your employer's competitor. Taking an employer's intellectual property for one's personal business is both a conflict of interest and theft. Most organizations also judge the appearance of a potential conflict to be a conflict. Many consider that co-authoring a paper with one's program manager constitutes a conflict. Finally, failure to disclose a potential conflict is generally considered misconduct.

Managing potential conflicts of interest generally requires full disclosure of financial or personal interests. Employees,[4] especially senior managers and all corporate officers, should sign disclosures. Any potential conflicts or appearance thereof should be acknowledged in writing as part of presentations or research reports.

Conflicts may be avoided in several ways: recusing oneself from decision-making, placing assets in blind trusts, obtaining employer permission for outside business activity. Not representing multiple parties to a dispute or deal seems like obvious advice. Not giving or receiving gifts of substantial value avoids more problems than just conflicts of interest. More controversial are "no dating" policies and anti-nepotism rules such as not supervising close relatives.

INSTITUTIONAL ETHICS

Beyond ethics-based policies written in the employees' handbook, managers in public institutions have an enhanced obligation to project an ethical vision to the public: "We are worthy of the public's trust. Our integrity is beyond reproach." The same standard should apply to any organization that accepts public funding. The explanation is easy. It is line management—from front-line to CEO—that creates and *models* an organization's culture of integrity and ethical behavior. Modeling ethical behavior creates a firm foundation for individual and shared accountability. It is the basis of thoughtful decision-making.

---

[4] Whether compensated or not.

When employees understand expected behavior by seeing it in action, modeled and projected by line managers, they derive a sense of common direction that applies to all staff, and they feel a sense of stability and continuity in a rapidly changing environment.

To the contrary when the CEO of a research organization or other top manager frequently takes long official trips with an administrative aide with little relevant formal training and who can add little identifiable technical or legal value for the organization, employees whisper among themselves and morale drops. If that aide gets an elevated title with an elevated salary based on having an associate's degree from a community college in a foreign country, the whispers only get louder.

Theorem: As a manager, your behavior is always on display.
Corollary: The employees see everything.

Topic for introspection or discussion: If scientific misconduct can destroy one's career, why do people do it? Would it occur less frequently if organizations had clear rules of conduct versus "Shared Guiding Values?"

Workforce Management

11

Effective workforce management must be a major ingredient in the success of any research enterprise. This chapter explores the aspects of operations that are most vital to executing the strategic vision and plans of the enterprise. In choosing one's team, both (1) recruiting, developing, and retaining employees in the context of the enterprise strategy and (2) succession planning are crucial to the continuing vitality of the enterprise. Closely related to—but not identical with—managing employee performance is choosing to compensate employees consistent with their contribution to the enterprise.

BUILDING YOUR WORKFORCE STRATEGICALLY

Whenever managers have opportunities to hire employees, they may be tempted to select and hire the "golden solution to yesterday's challenges" or to satisfy a presently perceived need for staff to meet a current crisis. Some searches aim to replace the employee who just retired or who was just lured away by a competitor. Others are conducted so narrowly that only a single pre-selected individual can meet the posted qualifications. In practice, each of these tactical ploys has some justification, but tactical hires may be inconsistent with the long-term health and strategic plans of the enterprise. Whether or not hiring managers currently have sufficient business to warrant a hire, they should maintain a strategic hiring plan for their business unit. A useful strategy for moving forward implies having an accurate skill-based inventory of one's present staff and its match to the unit's present and projected resource needs.

Without doubt, the staffing level plus support with contract labor should be sufficient to produce contracted deliverables on schedule. Therefore, the staff mix must be sufficiently deep in critical skills; otherwise a manager courts a single-point failure. Any new hires should adjust the staff mix to realize strategic goals and to be consistent with the positioning of the enterprise. As funds to hire additional staff are not always available, managers should enhance staff flexibility to respond to changing goals. Every manager should also remain aware of leadership potential of his/her employees for the purpose of succession planning.

Research organizations frequently hire term employees and post-doctoral fellows to meet short-term needs and to bring new blood into the fold. Not every one of these persons can or should be converted to regular (indeterminate term) status in the organization; the manager needs to maintain a healthy turnover of this group of term employees.

Newly appointed managers often object, "I inherited my group; I have to do my best with what I've got." These managers are correct ... at least for a while. Doing the best with the staff on hand demands that assignments are driven by deliverables rather than by staff preferences. That can be done by matching the matrix of staff skills with the matrix of skills needed for the deliverables. That task will be relatively easy if one assigns staff efforts by using a work breakdown structure, by making activity-based estimates, and by tightly tracking employee performance. Unfortunately, getting the best effort from all does not necessarily imply that all can meet their performance expectations.

Both managing out low performers and expanding the organization's business create opportunities for hiring new staff. As some new business may not be the basis of a long-term customer relationship, cross-training of existing staff can bring needed flexibility to a work unit. Still, the residual benefit of the short-term business should exceed any "infrastructure debt" that may be incurred. Some national laboratory managers object with a common mantra, "We don't do job-shopping" or "Our plate is already full." If the new business allows acquiring new talent or cross-training existing staff to handle new research or design tasks, the organization comes out ahead with clear residual benefit. If one's plate is full, consider getting a platter. In the end, one should manage staff levels and mix by the interests of the enterprise and not by banal slogans.

During most times in most organizations, opportunities to make a new hire are infrequent. Therefore, managers should treat every new hire as a precious opportunity that needs to be used wisely. Consider whether your unit's demographics are healthy or are marching over the hill, whether a large fraction will retire in a few years. Trying to balance the numbers of early-career, mid-career, and senior staff can present thorny choices. Should one hire an established mid-level performer who meets present needs or an early-career researcher with superstar potential? The skills matrix should indicate whether your unit is understaffed or weak in a critical skill or whether one area of expertise is more than the business needs. Sometimes a two-step process— move an employee to a new assignment and backfill—can reap double benefits.

HIRING THE BEST

If the enterprise seeks a wedge into a new business area, recruiting an internationally recognized expert puts the organization "on the map." However, the hiring manger should recognize that this person will need immediate staff to support his/her efforts. Senior managers should consider whether the mix of staff supports the positioning of the enterprise? If not, new hires can adjust the profile of staff capabilities as suggested in Table 11.1.

Each new hire may open new opportunities for existing staff. In a scientific organization, the manager should be looking to free as much as 10%–20% time of the professional staff for creative exploration. A large organization should have at least one guru (chief scientist, chief technologist) who is paid out of management overhead.

> **Theorem 1:** The most important thing that you do as a manager is to hire new career staff.
> **Corollary:** If you make a bad choice, the whole organization suffers.
> **Lemma:** You suffer the most.

> **Theorem 2:** Take a direct, active role in the hire of every career subordinate.
> **Corollary:** If you must terminate the employee, you will be more directly active than you'd like.

Job postings: The *position description* for any search must be accurate and comprehensive. The job content includes a descriptive job category, level, and title; it describes core job duties, the associated level of responsibility, and the essential performance expectations. The minimum required technical *knowledge, skills, and abilities* (KSA) should describe the minimum breadth of experience and knowledge and the extent of that experience. Minimum functional skills describe the requisite experience in managing and organizing people, activities, or information and the level of demonstrated communications skills. The description may include preferred characteristics that may serve as tiebreakers.

TABLE 11.1   Match the staff mix to enterprise positioning in making new hires

ENTERPRISE POSITIONING	MATCHING ATTRIBUTE
Technology leader	Find "the best and the brightest"
Most flexible	Expand breadth of staff
Market leader	Add project leaders and "rainmakers"
Highest quality	Emphasize candidate's track record
Lowest price	Prefer post-docs and students

U.S. law requires that hiring decisions be based on *essential* requirements of the position. If requirements are posted, they are generally presumed to be valid. List any physical and mental requirements of the position, for example, travel, hours or work on weekends. Don't write "wired" postings; instead, request a waiver of posting from HR. Avoid words that may imply discriminatory[1] hiring such as "young," "recent graduate," or "degree *required*." There is one exception: for a post-doctoral position, a PhD can and should be required.

The hiring process should be formalized and conducted using the same type of funneling process as used to select new products. Carefully documenting procedures prior to beginning a search is considered evidence of a fair process. The first stage of funneling is casting a wide net to create an applicant pool that can be screened to form a "short list" of several candidates. Already at the stage of the screening of initial responses, the hiring manager and search committee (if used) should determine whether the best target audiences have been reached and if the pool is sufficiently diverse in accordance with institutional policy. Screening the pool and then creating the short list of candidates yields the list of candidates to be interviewed and asked for references.

All interviews should demonstrate professionalism of the enterprise. The interviews have two purposes: (1) to collect information about the candidate and (2) to increase the applicant's interest in the position. By uncovering a candidate's abilities, talents, strengths, and weaknesses, the interviewers hope to determine how qualified the applicant is for the position and how well the applicant would perform in the job. Other important questions are more difficult to assess: Does the applicant fit with the unit? What is the potential of the candidate? Before the hiring manager does his/her interview, the other interviewers should have a feeling of whether the applicant is truly interested.

With respect to increasing the applicant's interest in the position—a crucial ingredient in recruitment—interviewers should determine the candidate's expectations of the challenges and opportunities presented by the position and the organization as a whole. They should address the candidate's concerns and answer questions—referring questions about any corporate policies and benefits to HR, which can answer authoritatively. The interview with the hiring manager should clarify the new employee's roles and responsibilities. With respect to recruiting a highly desired candidate, the hiring manager's responsibility is to sell the candidate on the organization.

A word of caution about written interview notes: The interviewer should limit any written notes, judgments, and commentary to those areas in which s/he has professional competence. For example, one might write "John knows his quantum physics" rather than "John is bright." Preparing formal memos

---

[1] Such avoidance is required by law in the U.S.

"to the file" immediately after the interview avoids failure of recollection if there is an appeal, law suit, or an audit of the interview process by a governmental agency. Those memos are the *only* material that should be shared with others. Read what is written in the memo as a judge or jury would, and destroy any informal notes immediately (as soon as the formal memo is prepared). When in doubt about the appropriateness of any given comment, strike it, rephrase carefully, and get expert advice.

SUCCESSION PLANNING

Theorem 1: Life is unpredictable.
    Corollary 1: No one is indispensable.
    Corollary 2: Everyone has a price.

Theorem 2: Your best people will be the first to want to move up or move on.

Senior managers should assure that every leader in their organization has at least one understudy or successor. To attract, develop, and retain the "best and the brightest"—employees who can and will make the greatest contribution to future success of the enterprise as its leaders—executives should *position* the organization as technology or science leaders. Hiring the best is only the first step in getting those with the most potential to actually deliver star performance. Therefore, the relevant executive must have an accurate, current *assessment of the performance* of each employee with respect to others in the work unit. In scientific organizations, the distribution of performance is multimodal as illustrated in Figure 11.1.

An assumption inherent in the figure is that the manager who is at least one step above the supervising manager has developed a scored evaluation system by which employees in the unit are ranked with respect to their contribution to

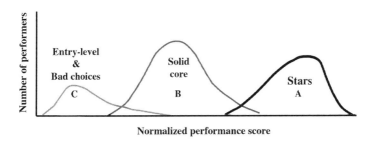

FIGURE 11.1 The distribution of performance is multi-modal.

that enterprise over which the manager has cognizance. The formal evaluation system must conform to the law and have documented, uniform performance measures and standards. It should have the following characteristics:

1. Orienting the employee to organizational objectives
2. Translating corporate goals into individual performance goals
3. Differentiating between levels of employee performance
4. Strengthening manager–employee communications
5. Evaluating performance based on specific actions and consequences
6. Linking performance to compensation and other personal actions directly, efficiently, and objectively.

Group C in Figure 11.1 is very much a mixed bag. It contains the newest hires of early career staff—for example recent post-docs—as well as staff whose performance is well below the expectations that were placed on them at the time of their hire. For the early-career employees, the manager should be assessing in every performance review:

1. What is the career potential for this hire and what are the metrics to track his/her progress?
2. Who will be his/her mentor and for whom is s/he the understudy?

Any organization that aspires to the position of a technology leader must have policies and procedures to manage out bad hires. However, these policies are worthless unless managers are trained to use them and have the courage to use them.

The bottom-up view of tracking post-entry-level staff starts with the evaluation of whether the employee's performance is making expected progress up the scale of value to the enterprise. Relevant questions to ask at least annually are "is this person a star in the making?", "what leadership level can this person handle?", and "what level of resources can this person manage?" The answers to these questions should generate two additional queries for the manager to ponder and act upon, "what training should s/he receive?" and "what stretch assignments are possible?"

Some employees will stop progressing as they reach the middle of the "solid core." Those who never get past the overlap region of Groups C and B should be candidates for managing out. That task is much easier in the U.S. in which almost all employees except government workers are "at will" *employees*.[2] In countries where a job is treated as a property right, an employee can only

---

[2] An at-will employee may be terminated for any reason except a reason that is illegal.

be terminated for just cause. In those countries, first hires are best done for a specified term; hiring beyond term positions must be done extremely carefully.

What drives the future of the enterprise is retention and future development of Group A "stars." Such employees should be given assignments that maximize their level of achievement, emphasizing creativity for the enterprise. The best of the best will need extensive resources—especially experimentalists—to become world leaders. Consequently the cognizant executive needs to make strong, visible efforts to secure resources for them and to tie their creative products to the strategic future of the enterprise. As these employees will always be prime targets for competitors to steal, management should make it highly undesirable for the star to leave.

Any employee above the midpoint of Group B should be evaluated with respect for the potential for a position two levels above present job assignment regardless of whether those steps are along the intellectual leadership axis or along the managerial axis. Employees can certainly exercise strong leadership of teams and task groups without having formal supervisory responsibilities. In such an evaluation, the manager should match the challenges of the higher position against the demonstrated management or leadership accomplishments of the employee. A manager who knows his/her employees will not be surprised that they lack a key characteristic for advancement. Develop those characteristics critical for success in the more responsible position via stretch assignments to develop the employee's knowledge, skills, and abilities.

By reviewing the development plans for the staff at least annually, one can continually build and evaluate employee accomplishments and potential, paying attention to more than technical skills. One can develop executive skills of employees as a deputy manager via term assignments that hone their judgment and expand their perspectives. Rotating assignments to the staff of a senior executive can also yield similar benefits. However, to develop and deepen management accomplishments, experience with line management responsibilities focused on deliverables is essential. Gaining supervisory experience can derive from assignments to multi-discipline teams or in matrix assignments. The executive's task is to get the best out of the scientific, technical, and administrative staff while building the credibility of the understudy—whether or not explicitly designated—in the promotional position.

By developing two successors for each management position, the senior executive maximizes his/her flexibility, maintains options in changing business environment, and increases readiness for business expansion. Having potential successors in line builds organizational agility and responsiveness to opportunity. Recognition by employees of the two-successor practice encourages and channels the competitiveness of the best staff and increases the organization's ability to compete in collaborative ventures. It also avoids the manager being held hostage by her/his staff.

To summarize, focus on retaining star staff. By envisioning a time scale for organizational change, you maintain opportunities for your most productive staff. You want options, and your best people want options. Avoid organizational arthritis in which condition every movement hurts. Manage out the least productive staff; this creates the resources and opportunities for solid and star performers. Keep room in the organization for post-docs and term appointees; they provide fresh blood flowing through the enterprise.

SALARY MANAGEMENT

The management of salaries for scientists and engineers seems to be an arcane topic in most research organizations and not much better understood by most managers than by the employees. This section briefly reviews a recommended practice used by the author. The assumptions are that (1) each employee receives a detailed performance evaluation at least annually based on the written input of the employee and the supervising manager; (2) the evaluation process is transparent to the employee and is legal, discoverable, and auditable; and (3) the evaluation is the basis for a ranking of employees.

Some organizations still use a simple raise management system in which an employee's assigned summary "grade" on his/her evaluation determines a percentage raise. This system has gross shortcomings[3] in that (1) it presupposes that the employee's salary is already aligned with performance, (2) ratings do not account for complexity, risk, relevance, or importance of the employee's work during the evaluation period, and, (3) it discounts the value of scarce special skills and knowledge. It is only consistent with quantitative salary package constraints if the distribution of grades is tightly regulated by senior management, there is no grade inflation, and top contributors asymptote to a score of "meets expectations."

Recognizing the obvious flaws of the raise management model, many organizations move to a modified raise management model tied to overlapping salary bands. In that model, the ranges of salary bands are based on averaged surveys of comparable labor markets. However, this model leaves unanswered critical questions: How are salaries set within the range? What should the distribution of salaries be within the ranges? In other words, the model leaves the foundational questions of any salary system unanswered: What should the distribution function of salaries be? And how is that distribution linked to value of the employee to the enterprise?

---

[3] Even worse is the system in which all employees get the same cost-of-living salary adjustment and nothing additional.

To have the full acceptance of its employees—and to function within the law—the enterprise needs to apply defensible rules for *pay equity*. One defensible principle is "we pay for performance and for contribution to the enterprise." Such a principle is not simple to apply equitably absent a ranking of employees. The second principle is easier to apply: "We do *not* pay for age, status, mere responsibility, or mere potential." Application of both principles should generate the relative ranking of employees in a manner that accounts for the quality and quantity of work, and for the complexity, risk, and importance of said work. It should not ignore the relevant skills, knowledge, and abilities that the employee possesses in reserve when the enterprise needs them in a work assignment.

As it measures the value of employee contributions to the organization and is an input to setting salary, a ranking system is legally discoverable in litigation; hence, it must be documented competently and carefully. Its output is a set of ordered scores that must be defensible "in court." The system should reflect institutional values and be based upon published criteria and on well-written documents that management can defend. It should explicitly include employee input and be generally consistent from year to year. As the manager who signs the employee evaluation, you "own"—must be able to defend—the ranking.

The question still remains: "How much should employees be paid?" The structure of salaries of scientists and engineers is a topic to be decided by the top management in consultation with the senior management team. That structure is deeply connected with the decision of top management's choice of positioning of the enterprise. Positioning tells an executive how to weigh the various categories of work activities in assessing the overall contribution of employees to the enterprise. Therefore, as management's decisions about positioning change, the concomitant changes will be those in the contribution assessment algorithm and not necessarily in the absolute pay scale. Figure 11.2 shows an actual salary scale versus normalized contribution (reward function) that was used for both scientists and engineers at a major national research and engineering laboratory with several thousand employees; senior managers are omitted from this graph. The key feature of the reward function is its nearly linear behavior for the entire core performer group. The poorest performers have considerably depressed salaries and are identified for being managed out (either dismissed or helped to find another workplace more in keeping with their abilities). The best performers are aggressively rewarded on a non-linear scale.

As special circumstances can lead to an extraordinary performance in a given year, the organization tries to separate out and reward the special performance with a bonus rather than with an inappropriately large increase to the base salary. As the value of the performance under average circumstances can vary from year to year, the functional line should be considered a band with a

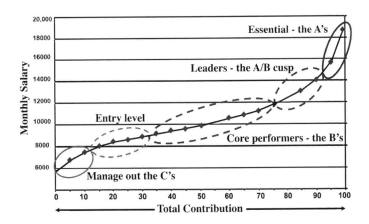

FIGURE 11.2   Salaries aggressively reward high performance by scientists and engineers.

full width of a few percent. For the core performers, the spread should be small at a given performance level and subject to correction within 3 years given the currently typical size of financial packages that managers have to administer. In that case, lump sum merit increases become infrequent corrections directly attributable to special circumstances, thereby providing additional incentives for extraordinary performance when possible. The reward function should be tied to relevant national or global salary markets, recognizing that the best performers are recruited worldwide. Using market peg points at the 20th, 50th, and 80th percentiles of performance should be satisfactory for scaling the reward function.

Some simple mathematical manipulation of the reward curve allows one to tie that structure to recruiting, developing, and retaining. That is seen most easily in Figure 11.3.

Flipping the reward function as shown yields the probability distribution function for the organization. Differentiating the probability distribution function with respect to the salary variable yields the ideal distribution of salaries in the organization. The lower left panel of Figure 11.3 shows the actual distribution of salaries in the author's research division after implementing this salary management approach for 4 years.

The salary distribution curve is linked directly to the strategic view of recruiting, developing, and retaining as illustrated in lower right panel of Figure 11.3. Poor performers see from portion (a) that there is little financial incentive to remain in the organization. Attracting the most promising early-career performers requires positioning them for rapid salary growth—segment (b). The salary range of segment (c) aims to incentivize a high level

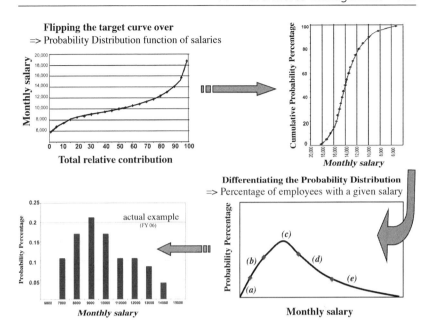

FIGURE 11.3 Manipulation of the reward function to obtain the probability versus salary.

of performance by the core staff. Retaining performers on the a–b cusp requires that they see a broad promise of potential financial reward—segment (d). Obviously hiring established world leaders to create the future for the enterprise requires the strong financial incentives of segment (e).

Other incentives, especially reserving time for high performers to be creative, opportunities for exposure among peers, and professional status all matter to creative scientists and engineers. Scientists may also say, "We're not in it for the money." BUT money does matter.

Managing Operating Risks

12

CONTROLLING RISKS DURING EXECUTION

In research enterprises, most programs and projects, by their very nature, push the envelope of what is known and what is possible. As a consequence, encountering technical difficulties along the way is almost guaranteed. Solving them requires managing scientists and engineers with relevant technical expertise as well as the courage to control the inevitable risks associated with operating at the boundary of the known and unknown. Three categories of technical challenges are the most common: (1) Solving many small difficulties takes more time than was planned or foreseen. (2) A major failure or underperformance occurs. (3) The responsible employee, team, or vendor responsible is unable to meet the performance specifications.

Once any such difficulties are identified and characterized, the research manager must be able to meet the challenges and take prompt action; delays only make matters worse. Among the investigative tools available to the manager are meetings, reports, and reviews. However, these tools have the potential to mislead unless they probe deeply enough. Especially in managing risks that have potentially severe consequences, post-hoc action is insufficient. Careful design of activities to prevent one-point failures, active surveillance of progress, collection of leading indicators, and rigorous control of all changes to program (project) plans or specifications are all essential. Using third-party experts is advisable both in risk management and in program reviews. Most *program reviews* are designed to highlight progress, areas for improvement, blind spots, oversights, and programmatic risks. However, with a sufficient number of viewgraphs, even serious flaws in hardware or in process design can be hidden. In reviews to mitigate risks, the use of viewgraphs or computer presentations is strongly discouraged, as such presentations can easily hide flaws in design or execution.

> **Theorem:** To ensure the quality of design or production, conduct engineering and process reviews as *in-depth engineering audits.*
>
> **Corollary:** Always conduct an engineering audit of designs before fabrication or production begins.

Whenever *corrective actions* are necessary, carefully assess the impact of the corrective actions on all aspects of the program (project); subsidiary, unanticipated consequences are always possible. Then, revise the program plans to reflect the impacts of any major changes on overall performance specifications, schedules, and resource requirements. The customers—or agencies—funding the work should receive written copies of the revised plans; surprises are never welcome. Finally direct special attention toward monitoring corrective actions taken to ensure that they actually resolve the problem(s).

As any sports referee knows, the calls should "sell themselves." If you must "socialize" every decision, your actions are not convincing in themselves. Therefore, in notifying sponsors of the work, do not confuse respectful pre-information with justifying your actions. Even then, consider how your words will be interpreted; as the basketball coach Red Auerbach said, "It's not what you say, it's what they hear." Likewise, consider how your employees will interpret your actions; they evaluate you every day. In *root-cause analysis*, once an incident occurs, investigators find that human error is a frequent root cause. Employees always suspect that the "blame game" will be at work; therefore, always insist on fair processes.

> **Theorem:** To do the right thing, you can't dismiss politics, and you can't be afraid to "play hardball."

Managing product quality, worker safety, information security, and compliance with environmental regulations require the active and visible engagement of all levels of line management of the enterprise. Their leadership by example not only demonstrates top-to-bottom commitment of the enterprise but also satisfies the *fiduciary responsibilities*[1] of corporate officers and their subordinate managers. Top management needs to reinforce accountability by means of pay, position, and promotion.

Without the visible engagement of line management, expecting employee commitment to corporate values of quality, safety, security, and environmental respect is likely to be a vain hope. Moreover, implementation through the line chain of command facilitates aggressive self-assessment and performance

---

[1] See Chapter 13.

TABLE 12.1 The correspondence between enterprise assets and risks

ENTERPRISE ASSETS	RISKS TO THESE ASSETS
Capital equipment	Accidents, obsolescence, misuse, vandalism, theft
Inventory and supplies	Theft, damage, waste
Staff	Accidents, degraded skills, lost morale, departure
Information, software	Misuse, corruption, theft, compromise
Reputation	Poor quality, late delivery

improvement programs plus consistency in approach, training, audit, and validation.

Some enterprises employ staff organizations to enforce the mitigation of everyday operational risks in what are effectively policing operations. While staff organizations can provide valuable subject-matter expertise as partners of line management, placing the full burden on them rather than on line management to enforce standards of behavior fosters an unhealthy "us-versus-them" culture rather than one in which all employees take responsibility for quality, safety, and environmental respect.

Some of the most common assets along with the corresponding risks that threaten them are listed in Table 12.1. In the final analysis, the line management from the top down is responsible for stewardship of enterprise assets. Emphasizing that point, the late, former SLAC Laboratory Director Burton Richter used to advise,[2] "It doesn't matter who screws up; it's the boss's fault." More specifically, Richter's advice means that prudent risk management avoids loss of total economic value and legal standing of both the corporation and its managers. In some instances, the losses can lead to civil and criminal penalties and liabilities to shareholder and customer suits. Failure to mitigate risks can lead to direct financial losses, loss of market share, and for publically traded companies, a drop in stock price.

A noteworthy management practice is control implemented via integrated risk management (IRM) systems. Figure 12.1 illustrates the top-down flow of control from top management of the enterprise. Feedbacks are provided by annual or semi-annual reviews and internal audits.

At the working level, this general scheme is depicted more simply in the diagrams of Figure 12.2 as applied to safety management but is easily adaptable to quality control or security management. At the center of the cycle of continual improvement of risk management practices is the first line of management—the work supervisor. The general scheme of panel (a) is made explicit in panel (b) for the specific case of managing workplace safety. One may also consider the supervisor at the center as a stand-in for the entire management chain of command.

---

[2] Private communication.

FIGURE 12.1 Top-down management of enterprise risks with appropriate feedback loops.

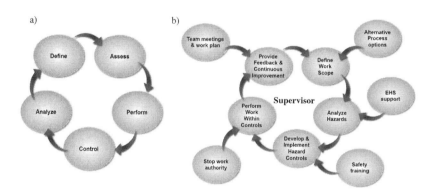

FIGURE 12.2 Panel (a) shows the general scheme of integrated risk management. Panel (b) specifies the components of the scheme for safety management.

Panel (b) also indicates how the staff support organizations—in this case the Environment, Health, and Safety (EHS) organization collaborates with the line management organization to improve safety practices.

Safety-related expectations of first-line supervisors are (1) analyze and plan the work; (2) identify and review all relevant hazards and applicable controls; (3) determine all relevant authorization requirements; (4) obtain appropriate authorization; (5) prepare required documentation of the work and its authorization; and (6) ensure that the authorizations are approved before commencing work. If there are any planned changes to work scope or hazards, the first-line supervisor should inform the chain of command and the divisional or departmental safety coordinators.

> **Axiom:** You can't be everywhere.
> **Theorem:** You can't be everywhere at all times.

Even the procedures within IRM are insufficient to ensure zero serious accidents in the workplace. From the theorem, it follows that every employee must be empowered to serve as the eyes, ears, and voice of the supervisory chain when it comes to workplace safety. Delegating *stop-work authority* to all employees accomplishes that empowerment. To be effective stop-work authority demands the constant and visible encouragement and support by all line management.

> All personnel are authorized to stop any work activities that appear to present
> an imminent danger, regardless of status of persons performing work.

Executives—including senior management—must exercise adequate, ongoing oversight of work activities. IRM also requires active engagement of executives to take prophylactic measures to mitigate risks to safety, security, and quality:

1. Conduct unannounced walkthrough inspections of all workspaces at least annually.
2. Report safety, security, and quality concerns at least one step up the chain of command.
3. Maintain safe, secure, and orderly work areas; identify and remove unused equipment from active work areas to storage areas as practical.
4. Provide safety, security, and quality orientation and training to newly assigned personnel; evaluate training needs of assigned personnel whenever their job duties change; evaluate employee EHS and security performance during their annual review.
5. Take responsibility for the safety, security, and quality of any contracted work; assure qualified (and cleared, if necessary) contractors are selected, hazards are identified, and work is performed safely and securely.

Safety and security records cannot be improved based only on tallies of rare events. Days between security incidents and violations or between accidents and accidents per worker-hour are all *lagging indicators* that are not under a supervisor's *direct* control. Chance circumstance has a large effect. Although those records are important to maintain, far more effective at mitigating risks is management actively driving behaviors and practices that promote safety, security, and quality by developing safe, secure, and high-quality work habits in employees to stay out of harm's way. Just as lagging indicators are tallied, management should record the proactive measures that it does control as leading indicators. In the event of incidents, the records will indicate prudent, rather than negligent, supervisory behavior.

Despite all precautions, an incident may happen. In that case, ensure any incidents involving assigned personnel, whether on-site or off-site during official travel, are promptly reported up the chain of command. Participate in reviews of any occurrences involving assigned personnel. Complete reports of both proximate and root-cause analysis promptly and accurately, and identify and implement appropriate corrective actions.

Whether for safety, security, or product quality, the chain of command must consciously build excellence through the disciplined improvement of established systems. That process begins by assigning single-point accountability for all established systems with well-articulated and documented standards. It proceeds through both self-assessment and external reviews designed to derive lessons learned and recommend modification of procedures to improve and enhance systems. Continuing processes improvement is a never-ending obligation of top management and all subordinates.

Theorem: Yes. You'll have to fight for time to do your own research!

Structures and Governance 13

The first chapters of this text introduced the concept of multiple networks of externalities—business, societal, and professional—that provide contexts for the research enterprise. Those networks have connection points with the corporate enterprise, with top management, and with employees. Between the top executives and the workers are managers in the middle, structured into internal networks not yet discussed. It is time to open that box. In most of what follows concerning internal organizational structure, the reader may substitute the words Project, Program, and Product Line interchangeably. The section concerning governance assumes that the enterprise is structured legally as a corporation.

Three concepts are of central importance when discussing organizations: (1) *Authority* is the legitimate, hierarchical right to issue orders, which others must follow. Authority can be delegated downward to lower levels in the organization; however, responsibility always rests with the individual. (2) *Responsibility* is the obligation to effectively perform assignments incurred by individuals due to their roles in the formal organization. (3) *Accountability* refers to the state of being totally answerable for the satisfactory completion of a specific assignment.

With the exception of small partnerships, the lines of authority flow in a specified, hierarchical manner through a number of levels that depend on the size of the organization and the preferences of the Board of Governors (Directors). The *flow of authority* is typically represented in an organogram (or organization chart). An organization may have multiple organograms to depict reporting relationships, authority to commit funds internally, accrual of costs and profits, signature authority to commit (legally bind) the enterprise, coordination responsibilities, and so forth.

ORGANIZATIONAL STRUCTURES

An Internet search will reveal many sorts of organograms, some of which are shown in Figure 13.1. In studying these models, the reader should ask, "Who can commit funds?" and "Who can commit people?" When these are not the same persons, a potential for internal conflict and external embarrassment with customers exists and must be managed by senior or top executives.

In *functional organizations* (Figure 13.1a), employees are organized into units according to their skills—for example, mechanical engineering, electrical engineering, computations, physics, chemistry, etc. Functional structures have the advantage of creating large teams of highly skilled specialists grouped by competence, thereby lowering personnel costs through reduced duplication of staff. In some business settings, communication and coordination problems may be lessened. The price to pay for these benefits is the difficulty of crossing functional lines. Yet many, if not most, projects require a mix of skills, hence necessitating multiple negotiations. The strong silos of

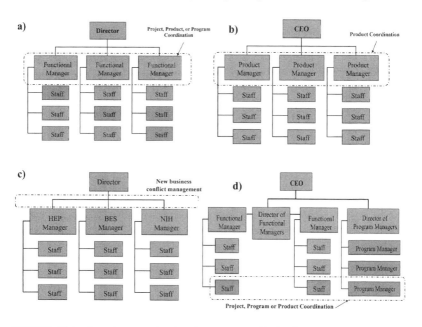

FIGURE 13.1   Popular organizational structures: (a) the functional organization, (b) the product line or program organization, (c) the customer-oriented organization, and (d) a strong matrix organization. The agents of coordination are indicated within the dotted lines.

the functional system create barriers to horizontal flow of information, which can result in slower decision-making. The silos also produce managers with narrow work experience and with a conflict of interest in servicing project or products that originate outside their respective silos. Moreover, the functional emphasis generates loyalties to one's brother or sister engineer rather than to delivering the product; this can impede project completion. Finally, if the enterprise is a service provider of skills, this organization is just a product line organization with a different kind of product. Note that resolution of conflicts is at the level of functional managers. The program development (marketing) manager, who brought in the money for a task, has extremely limited power to enforce actions in which the quality and timeliness of the product is of primary importance.

A quite different structure used by many companies is the *product line organization* of Figure 13.1b. The product line could be hardware or hard goods made by the organization. Alternatively it could be a multi-year program of research, or it could be a project. In the case of a project, the product line has a finite, specified duration that introduces special management challenges of its own. The multi-billion dollar project of building the National Ignition Facility[1] in the U.S.—the world's largest high-energy laser—was managed as a product line organization.

Managers in product line structures have the advantage of a specialization enhanced by broader experiences outside of just one discipline. As the structure is designed to produce deliverables to meet customer needs, assessment performance of work units can be based on the quality, timeliness and cost of product, and the level of customer satisfaction. Moreover, the confusion between cost centers and profit centers that sometimes bedevils functional organizations is avoided. Hence, attribution of gross income delivered to the corporation is clearer.

Because the product line managers own all their resources—in particular, personnel and contracts for services—decision-making can be faster. In bringing in new business within a product line, the line manager controls and is empowered to commit his/her resources. That fact alone makes marketing more convincing and makes closing a deal with a customer far less cumbersome. An obvious disadvantage of the product line structure is that duplication of personnel resources often increases costs. The most frequently used way around that drawback is having some work done on contract by a company that sells services as its product. A further disadvantage is that product lines are meant to be distinctive; hence, it can be awkward to coordinate across departments.

---

[1] https://en.wikipedia.org/wiki/National_Ignition_Facility.

A modification of the product line structure seeks to sharpen the focus on customer satisfaction while limiting conflict between product line managers who want to sell to the same customer. The *customer-line* organization of Figure 13.1c coordinates the marketing activities of the enterprise by applying the "one hand in one pot" rule. Such a scheme is frequently found in U.S. national laboratories. Advantages of the customer-line structure are a strong focus on customer needs as well as the creation and delivery of products and services tailored to specific customer needs and preferences. The disadvantages are the same as those of the product line organization augmented by the potential to make decisions that pander to the customer but may hurt the enterprise in the long run.

A well-known approach that aims to maximize advantages of both the functional and the product line systems while minimizing their respective disadvantages is the strong *matrix organization* of Figure 13.1d. In matrix structures, individuals from functional (home) units are assigned on temporary basis to the program manager (host) who also manages a product line (host) unit with its own specialized employees. The "matrixed" employees work in a multiple-command system with two supervisors alongside employees from the host unit. This approach allows efficient management of large, complex tasks with a full complement of requisite employee skills. When the matrixed functional employees complete their assignment, they return to their home organization. Where programs rather than projects are the product line, the assignment of functional experts can be for an indefinite term. This system requires a high level of management skills and high levels of coordination between managers at an equivalent level in both the host (program) and home (functional) divisions. Careful plans and procedures—best specified in writing—are needed to minimize negative effects of dual reporting, such as the employee being trapped in the middle of two managers with disagreements or with the employee trying to play one supervisor off against the other.

The matrix structure has the advantages of visible objectives, efficient utilization of resources, and better coordination and information flow. The matrix also encourages functional managers to continually enhance the training of functional employees in their areas of expertise. For project assignments, the matrix assures employees of a home after project completion. Naturally, the host division must pay a premium (personnel burden) over direct charges (salary plus fringe benefits) for the right to send employees back to the home division. Many of the disadvantages arise from matrixed employees having multiple bosses. Matrix structures are more complex to control than functional or product line models because of the differing priorities of the program and functional managers. Inevitably some duplication of effort and some conflict remain.

In organizations, the *chain of command* (vertical line of authority) describes who reports to whom. The unity of command implies that an employee reports to only one boss; this unity is violated by matrix structures. Therefore, written rules for priority of orders of the two supervisors—one, home and one, host—are essential. *Span of control* refers to number of subordinates a supervisor can directly control. The more complex and variable the position requirements, the smaller that span should be. One also differentiates line versus staff authority. A manager with *line authority* can command immediate subordinates in his/her chain of command. S/he contributes directly to creating and marketing the organization's research or projects. In contrast, an executive with staff authority can advise, but not command, others. Instead, the activity of staff executives is to support line activities. Staff executives can have line authority over staff support units (for example, an office of public relations).

Delegation of authority entails passing authority to others (subordinates), although the ultimate *responsibility* remains with the original manager. Delegation extends "what a person can do" to "what a person can control," thereby releasing time for the superior manager to focus on more important or more time-sensitive work or on work that s/he is uniquely qualified to perform. In addition, delegation provides an opportunity to develop the initiative, skill, knowledge, and competence of subordinates. While competent and confident managers should have no difficulties in delegating tasks, several factors work to discourage delegation, even when it would appear advantageous to do so.

One common obstacle is a manager's preference for performing work as opposed to managing the work. That preference is often rationalized by a belief that "I can do it better myself" or "I can only accept perfection in this task." Insecurity or uncertainty regarding tasks and/or an inability to explain tasks can increase a manager's reluctance to delegate as can the lack of experienced or lack of competent subordinates. A more acceptable obstacle is the criticality of the decisions that the task entails plus a confusion regarding responsibility and authority. Regardless of his/her decision regarding delegation, the manager retains the responsibility that the task be discharged competently in a timely manner.

In problem situations in which the responsibility of the manager or executive is acute, some level of delegation can be helpful. The charge to the subordinate can be as minimal as investigating the situation: "Look into the problem and report all the facts to me; then I'll decide what to do." Or it could be far more extensive: "Look into the problem, let me know what you intend to do, but do not take action until I approve." To make the best decision in any given case, the manager must first analyze the job, plan the delegation, select the right person, and track the work to completion.

INSTITUTIONAL GOVERNANCE

Governance codifies the nature of flow of authorities, duties, and responsibilities and provides the superstructure that assures that the corporation has a competent, ethical, and effective CEO and subordinate officers who operate the organization effectively, legally, and ethically to produce value for owners (shareholders). It should seek to instill a culture of integrity, legal compliance, and accurate, independent auditing of enterprise finances. Governance mechanisms should ensure appropriate responses to the concerns of owners and fair, equitable dealing with stakeholders. When governance fails, governments often intercede with prosecutions, new legislation, and imposition of new liabilities for directors. The flow of authority and responsibility is depicted in Figure 13.2.

Roles of owners and shareholders: Shareholders of publicly held corporations are not typically involved in the day-to-day management of corporate operations. However, they elect representatives (directors) as guardians of their interests. Shareholders have the right to the information necessary to making prudent investment and voting decisions. If the organization (for example, a national lab) is owned by a government agency, but operated by a contractor, the agency personnel are not involved in day-to-day management of the operations (well, hardly ever, and every year micromanagement gets worse and worse). The government does choose the operating contractor, reserves the right to veto choice of top management and officers, and can withhold the

FIGURE 13.2   Flows of authority and responsibility though enterprise governance.

management fee if the annual performance of the contractor is unsatisfactory. If the enterprise is privately held, the owners often act as top management.

Roles and responsibilities of Board: The members of board of directors have a *fiduciary*[2] *responsibility* to oversee the performance of the corporation and its management on behalf of owners. That responsibility carries a duty of *due care* with respect to the *due diligence* standard.[3] Directors have a duty of loyalty that includes the good faith disclosure of potential *conflicts of interest*. Violations of loyalty include knowing transgression or conscious disregard of conflicts of interests for improper personal benefit. The prime task of the Board is to select, oversee, evaluate, and compensate a well-qualified and ethical CEO. The Board may remove the CEO for non-performance or other behavior inimical to the interests of the owners. The Board also approves the appointment of subordinate corporate officers and can remove them if necessary.

The Board is legally bound to monitor diligently business operations, guiding and approving the strategic plan and movement of the corporation. It reviews and approves significant corporate actions as required by law and oversees legal and ethical compliance of the enterprise. With respect to finances, its audit committee is responsible for the integrity of corporate audits and for the clarity and accuracy of the corporation's financial statements and reporting.

To discharge its responsibility for the continuity of business operations, the Board advises top management on significant corporate issues and reviews management's plans for business resiliency. It also plans for developing managers and for succession, in part by assuring the compatibility of the strategic plan with incentive programs for executive retention.

Roles and responsibilities of the CEO: The CEO is authorized to operate the enterprise in an effective and ethical manner, to run its day-to-day business operations, and to carry out its strategic objectives within an *annual operating plan*[4] and budget as approved by the Board. To carry out these responsibilities, the CEO is charged with creating an effective organizational structure staffed with qualified senior managers. As the chief risk officer of the enterprise, the CEO must identify and manage all varieties of operational risks, generally with the assistance of the corporation's general counsel, a chief security officer, a chief information officer, etc. The CEO is responsible for creating and executing the strategic plan of the enterprise, generally aided by a planning office.

---

[2] Generally, the Board's *fiduciary duties* are owed to the corporation rather directly to shareholders; however, duties to the shareholders are often enforced by the shareholders through a derivative suit against board members on behalf of the corporation.

[3] As specified by the law of the country of business registry or incorporation.

[4] Produced by the CEO together with a chief operating officer (COO).

Subordinate corporate officers: To carry out *fiduciary duties* on behalf of the CEO, the Board appoints corporate officers[5] and agents of the enterprise. Through a chief operating officer (COO) or a general manager, and with the fact-finding activities of an *Internal Audit* office, the CEO discharges responsibilities for the lawful, safe, health-conscious, and environmentally respectful operation of the enterprise. Preparing accurate and transparent financial reports and disclosures consistent with *cost accounting standards* is the task of the chief financial officer (CFO). Together the CEO and CFO are responsible for certifying accuracy and completeness of financial statements as well as for the effectiveness of internal controls and controls of financial disclosures. All corporate officers are expected to assure the ethical standards of the organization—leading by example—in full compliance with the letter and spirit of the law.

---

[5] Corporate officers are liable to corporation for violation of fiduciary duties in the same manner as directors.

Technology-Transfer Case Study

14

Universities and government-sponsored research organizations have many ways to transfer their research product to the commercial sector. In so doing, research can be developed into products (including knowledge) with practical utility and economic value. The transfer process aims to generate income; disseminate and diffuse the science or technology through the research, industrial, or consumer sector; and stimulate further efforts to develop innovative intellectual property. In addition, some funding agencies, for example in the United Kingdom, put very strong emphasis on long-term grant recipients to foster strong technology-transfer efforts based on their government-funded research.

Most descriptions of *Intellectual Property (IP)* include copyrights, trademarks, patents, and trade secrets. However, a very important form is technical know-how or skill-based information, the skill set in the possession of an organization's workers. While sharing similarities with trade secrets, know-how cannot be simply written down in a recipe even when the product remains within the same company. A characteristic of know-how is the manufacture of a product with quantifiable properties superior to those found in comparable products. The most effective means of transfer of know-how is through exchange of personnel and joint ventures between the research entity and a commercial entity to form a jointly owned commercial venture. The bulk of this chapter is devoted to a case study of such a transfer. But first, a more general explanation is in order.

A primary method of technology transfer for universities and government labs is the protection of their intellectual property through patents. The patent of an invention is a legal vehicle issued by a government that gives its owner the exclusive rights to market, sell, distribute, and make derivatives depending on its invention. Those exclusive rights last for a limited period (often 20 years) as specified in law. Once a technology or a design is patented, the rights to use or market the technology can be transferred to other entities via licenses, either exclusive or non-exclusive.

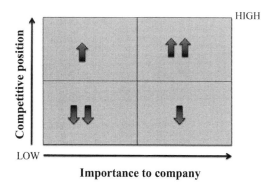

FIGURE 14.1   A company's technology-transfer decision matrix. Arrows indicate the advisability of investing in technology in that sector.

Most major universities have formal technology-transfer offices with the mission to solicit licensing agreements from the private sector. An impediment to such agreements is the long time that it may take to bring a technology to market especially in today's world in which *time to market* is crucial to commercial success. Looking deeper, one can understand that between the patent and the commercial product is a critical step of innovation and development. The steps of innovative engineering development followed by test and evaluation of the product are neither necessarily quick nor inexpensive. Therefore, the commercial entity considering to purchase a license must decide based on technological forecasting and market research that it has the financial tolerance for a period of negative cash flow.

The attractiveness of technology transfer depends on the competitive positioning of the *derivative products* (see Chapter 9) and the potential importance to the company of the derivative products. Figure 14.1 shows a possible decision matrix based on these considerations.

Other methods of technology-transfer include (1) cooperative or joint research agreements between the research entity and the commercial entity, (2) dissemination of information in conferences and tradeshows, and (3) the creation of spin-off commercial ventures.

KYMA CASE STUDY

In April 2005, Elettra-Sincrotrone Trieste S.C.p.A. kicked off its first workshop devoted to the building of FERMI, the world's first seeded free electron laser (FEL) user facility, making use of the recently retired linear accelerator (linac) injector of the Elettra storage ring light source. This FEL was to have two separate beamlines, each with sufficient sections of undulators[1] to bring the FEL to saturated operation at wavelengths as short as 4 nm.

Instead of hiring term contract engineers and technicians to build the undulators for FERMI, Sincrotrone Trieste decided to adopt an aggressive technology-transfer effort. Its idea was to establish a spin-off company formed by different legal entities, to develop a specific innovative business, under a unique central management and responsibility toward the customer. The new company would design, realize, and install all of the 18 undulators for FERMI@Elettra project. By the end of 2006, Sincrotrone Trieste issued an open European tender with the purpose of finding potential partners for building, measuring, and installing the undulators. Potential partners were required to set up a new company (NewCo) for that purpose, together with Sincrotrone Trieste. The capital of the company was formed by 51% of shares of the NewCo as intangible assets supplied by Elettra-Sincrotrone Trieste and the remaining 49% of liquid capital supplied by the partners: Cosylab (a Slovenian instrumentation company—27%) and Euromisure (and Italian electromechanical firm—22%).

Sincrotrone Trieste contributed by transferring its know-how concerning the design and building of undulators to NewCo. The knowledge transfer was financially evaluated and assigned a monetary value. With all partners fully committed, Kyma Srl was established in August 2007 and Kyma Tehnologija d.o.o. was established in Slovenia in July 2008. The business operations of Kyma started by end of 2008. Kyma delivered all 18 undulators for the FERMI@ Elettra project on specification, on time, and on budget. It then began to compete for supplying insertion devices to the global light source market in 2010. Since that time, Kyma has established contracts with major scientific institutions worldwide. The company has established a solid record of profitability with 2 M€ profit after taxation in fiscal years 2008–2013. The company is now in the process of considering how to build on past success in expanding its product line.

Kyma is organized in the form of a *virtual company*—an extended organization, which directly controls an interrelated set of coordinated processes carried out at different locations, by different legal entities (independent companies). Therefore, a virtual company is an unbounded organization, not delimited by the walls of its buildings but rather through its processes, which can move knowledge and information, goods, and services, with little regard to the physical place where they are produced or delivered. Key attributes of a virtual company are that it can be lean, extended, and adaptive to changing business conditions and to the evolution of its underlying technologies.

---

[1] Undulators are precision magneto-mechanical structures with an alternating array of north and south magnetic poles.

KEYS FOR LASTING SUCCESS OF TECHNOLOGY TRANSFER

The Kyma story illustrates the degree to which success begins with all partners being fully committed to the development of the business rather than on short-term profit. The spin-off concept assured a hot start with an important contract granted immediately for a state-of-technology product, thereby avoiding the long period of negative cash flow that can discourage investors. With a lean, extended, and adaptive virtual organization and clear positioning at the boundary of invention and innovation, the start-up was able to adopt a full customer-oriented approach based on an extreme focus on project and process management. The necessary ingredient for this virtual company approach to work is a strong trust relationship that is continually nurtured by the Kyma senior management. Figure 14.2 provides a schematic of the interplay of research organizations and the commercial sector.

Even a hot start can fail in the long run unless the start-up is strongly committed to a continual improvement of products and processes as well as to a

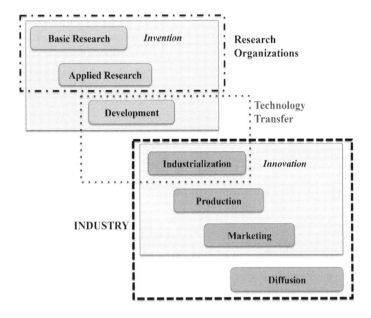

FIGURE 14.2 Interplay of research organizations with industry in the technology-transfer process.

carefully nurtured relationship with the vendors of its supply chain of specialty materials (in this case permanent magnet materials). A technology spin-off like Kyma thrives best if it develops close, continuing relationships both with the pure industrial and pure research environments. Doing so means keeping multiple lines of communication open to develop operative innovation partnerships even when immediate sales are not at hand.

Virtual organizations necessitate associations, federations, relations, agreements, and alliances, as they essentially are partnership networks of disseminated organizational entities or self-governing corporations. In principle, the "Prizes" of partnership networks are a wider range of products and services, increased know-how within the alliance, and consequently a better position in the market. The "Price" is continuing proactive communication and engagement that builds and nurtures trust relationships among allies.

STRATEGIC PARTNERS AND ALLIES

Successful spin-offs depend on rapidly fitting into a network of symbiotic relationships with strategic partners and allies, with complementors and suppliers feeding other parts of the value chain, with the relevant research community, and with natural customers. A *strategic partnership* is an agreement between two or more parties to share finances, skills, information and/or other resources in the pursuit of common goals. Such partnerships come in various forms: joint ventures for a defined timeframe over a specified market segment and strategic alliances that aim at strengthening competitive positions of the members of the alliance in the marketplace. The members of the partnership expect the consequent benefits to begin with new customers and increased sales. Over time, the partners hope to expand these benefits to access wider and more diversified markets. In themselves, the legally binding agreements can accelerate the acquisition of key assets (licensing rights, acquiring expertise, efficiencies, etc.), sharing business risks, and offering a variety of measures to devote more attention to customer satisfaction.

A step-by-step outline for creating an effective partnership begins with each potential partner determining what it wants from the partnership, that is, the objectives and the desired outcomes of signing a partnership agreement. With all parties understanding their interests and issues, a multi-party negotiation can begin. Fruitful negotiation presupposes careful strategic planning with key stakeholders both from relevant businesses and from cross-functional team members. The product of the negotiation should be unambiguous agreement on revenues and other benefits for each party. The parties should agree on a formal marketing plan with defined roles responsibilities, marketing programs, costs, and so forth. Naturally, the business perceptions of each of the parties will differ in some respects. Therefore, the parties can expect a rocky ride unless they have agreed on how to track and report measures of progress. Finally, it is time for the lawyers to put the agreement in writing; otherwise no one will be held accountable!

Realistically, however, these types of alliances often do better in the boardroom than "on the street." In practice, the challenge is most often the variation in *cultural* values within the different organizations rather than *technical* disagreements across the engineering and marketing staff members. Cultural discord easily leads to decisions out of sync, misinterpretation of events, and frustration among all parties. As already noted regarding the virtual company, the partners have to be completely confident of each other; strong trust relationships are the essential foundation. That confidence needs to be driven by line managers; the partners need strong management capabilities so that customers want to buy projects rather than products.

In summary, creation of a strategic partnership requires (1) careful planning and implementation; (2) precise definition of skills and competencies necessary for the whole business; and (3) processes, methods, and procedures used to develop products. Partnerships form most easily around a whole product in mid-value chain. Common business wisdom is to keep agreements simple. Once a wide network of partners and complementors forms, it will be easier to create or expand the market, for example, via conferences and trade shows. With large partners, it is best to work across peer levels of line managers; a small company courting a large firm is best advised to focus its energy to establish a relationship with the district office level of the larger firm.

A word of caution: as Robert Burns observed, "The best laid schemes o' mice an' men/Gang aft a-gley." To translate, do not be surprised to discover that the most difficult partner to manage is one's own company.

Recommended Resources 15

CHAPTER 1

Supplementary Materials: Chapter 1—Introduction. https://www.crcpress.com/
9780367255855

CHAPTER 2

Supplementary Materials: Chapter 2—ELM Space and the Network. https://www.
crcpress.com/9780367255855

See www.businessmanagementideas.com/leadership/top-4-theories-of-leadership/3351

Derue, D. S; Craig, J; Wellman, E; Humphrey, S. E (2011). "Trait and Behavioral
Theories of Leadership: An Integration and Meta-Analytic Test of
Their Relative Validity," *Personnel Psychology*, 64(1), 7–52 https://doi.
org/10.1111/j.1744-6570.2010.01201.x

Bolman-Deal Four Frame Model, www.slideshare.net/PhilVincent1/fourframe-
model, and *Reframing Organizations: Artistry, Choice, and Leadership*, Lee
Bolman and Terrance Deal, 6th edition, John Wiley and Sons (2017).

An example of a parody of the Blanchard model may be found in *The Mafia Manager:
A Guide to the Corporate Machiavelli* by V., IBSN 0312155743.

NASA Leadership Model Executive and NASA Leadership Model Manager www.
au.af.mil/au/awc/awcgate/nasa/leadership/executive.pdf and www.au.af.mil/au/
awc/awcgate/nasa/leadership/manager.pdf.

CHAPTER 3

www.nsf.gov/statistics/2018/nsb20181/report/sections/research-and-development-u-s-trends-and-international-comparisons/cross-national-comparisons-of-r-d-performance - country-and-regional-patterns-in-total-national-r-d.

Supplementary Materials: Chapter 3—The Research Environment. https://www.crcpress.com/9780367255855

CHAPTER 4

The reader is referred to an excellent summary of methods by David S. Walonick, "Overview of Forecasting Methods," www.statpac.com/research-papers/forecasting.htm.

CHAPTER 5 FOR INFORMATION ABOUT THE VARIANTS AND USE OF THE BALANCED SCORECARD, SEE

Kaplan, Robert S; Norton, D. P. (1992). "The Balanced Scorecard – Measures That Drive Performance." *Harvard Business Review* (January–February): 71–79.

See also www.balancedscorecard.org/Resources/About-the-Balanced-Scorecard.

Wikipedia offers fifty-seven references about the balanced scorecard method and its effectiveness as strategic management tool. https://en.wikipedia.org/wiki/Balanced_scorecard

Supplementary Materials: Chapter 5—Strategic Planning. https://www.crcpress.com/9780367255855

CHAPTER 6

Supplementary Materials: Chapter 6—Case Study: A Balanced Score Card. https://www.crcpress.com/9780367255855

Stephen Kerr, "On the folly of rewarding A while hoping for B." *Academy of Management Journal*, December 1975, volume18, number 4, p.769.

CHAPTER 7

Supplementary Materials: Chapter 7—Business Plans. https://www.crcpress.com/ 9780367255855

CHAPTER 8 COMMUNICATIONS FOR MANAGEMENT

Edward Hall's book *Beyond Culture* published in 1976 introduced the concept of high- and low-context cultures and its applications to cross-cultural interaction. There are many descriptions of Hall's approach on the Internet. As always Wikipedia is a good starting point which maintains active links: https://en.wikipedia.org/wiki/ High-context_and_low-context_cultures.

See MIT Sloan School 15.667-Negotiations & Conflict Management. https:// ocw.mit.edu/courses/sloan-school-of-management/15-667-negotiation-and-conflict-management-spring-2001/. Professor Rowe's *"A Butterfly's View of 15.667-Negotiations and Conflict Management"* is a helpful synopsis, http:// ocw.nur.ac.rw/OcwWeb/Sloan-School-of-Management/15-667Spring2001/ StudyMaterials/index.htm

The publisher Elsevier maintains a useful website with questions and answers related to permissions: www.elsevier.com/about/policies/copyright/permissions.

Supplementary Materials: Chapter 8—Communications 1: Meetings. https://www. crcpress.com/9780367255855

Supplementary Materials: Chapter 8—Communications 2: Writing. https://www. crcpress.com/9780367255855

CHAPTER 9 STRATEGIC MARKETING OF SCIENTIFIC ORGANIZATIONS

A comprehensive synoptic text is *Strategic Marketing Management*, by Alexander Chernev, Cerebellum Press; 9th ed. edition (January 1, 2018).

See also the following course notes available at MIT Open Courseware: MIT 15:810: Marketing Management; MIT 15.840: Marketing Management—M. Norton & D. Ariely; MIT 15:821 Marketing Strategy. http://ocw.nur.ac.rw/OcwWeb/Sloan-School-of-Management/index.htm

CHAPTER 10 RESEARCH ETHICS

Ethics of Emerging Technologies: *Scientific Facts and Moral Challenges*, by Thomas F. and Mariam D. Budinger.

"Ethics and the Welfare of the Physics Profession," Kate Kirby and Frances A. Houle, *Physics Today* **57**, 11, 42 (2004); https://doi.org/10.1063/1.1839376.

Federal Policy on Research Misconduct, www.ostp.gov/html/001207_3.html.

Supplementary Materials: Chapter 10—The Eightfold Way. https://www.crcpress.com/9780367255855

Supplementary Materials: Chapter 10—Research Ethics. https://www.crcpress.com/9780367255855

CHAPTER 11 WORKFORCE MANAGEMENT

Supplementary Materials: Chapter 11—Workforce Management. https://www.crcpress.com/9780367255855

Supplementary Materials: Chapter 11—Performance Management. https://www.crcpress.com/9780367255855

CHAPTER 12

Supplementary Materials: Chapter 12—Managing Operational Risks. https://www.crcpress.com/9780367255855

An important approach to managing safety is described in *Engineering a Safer World* by Prof. Nancy Leveson, MIT Press 2012. A pdf version is available free from the MIT Press at https://mitpress.mit.edu/books/engineering-safer-world.

Index

A

Accountability, 4, 97–98, 112, 116–117
Accounting, 52
Active listening, 76
Annual operating plan, 123
Authority, 117, 121–122
Authorship, 94–95

B

Balanced scorecard, 40–41
BATNA, 79
Board of directors, 123
Budgets, 53–54
Business environment, 2
Business opportunity, 61–63
Business plan, 59–68

C

Chain of command, 121
Change process (controlling), 40–41,
 45–46
Chief Executive Officer (CEO), 4, 123
Chief risk officer, 29, 123
Command presence, 12
Competitive advantage, 63–64, 66, 88–89
Conceptual design report, 72–75
Conflict management, 77–78, 80–81
Conflict of interest, 93, 95–96, 119, 123
Contingency budget, 34, 35, 47
Copyright, 71, 125
Corporate strategy, 42–44
Cost accounting standards, 57–58, 124
Cost-benefit analysis, 30
Cost-driver, 51
Cross-training, 100

D

Delegation, 121
Delphi technique, 34

Direct costs, 51
Due diligence, 123

E

Editing, 71
Editor-in-chief, 73, 75
ELM space, 10–12
Employee compensation, *see* Salary
 management
Ethical conflicts, 92–93
Ethics, 91–97
Executive summary, 65, 73, 75

F

Fair process, 92
Fair use doctrine, 72
Feature-benefit discriminators, 61,
 86–87
Fiduciary responsibilities, 112, 123–124
Financial plan, 49–51
Financial responsibilities, 56–58
Financial tools, 52–55
Fixed target paradigm, 32
Forecasting, 31–34, 126
Funneling, 62, 84–85, 102

G

Gap analysis (swot), 39, 60
Governance, 117–124

H

Hiring, 99–103

I

Indirect costs, 51
Information system, 48
Integrated assessment, 31
Integrated risk management, 113–115

Intellectual property, 89, 125
Internal audit office, 124
Interviewing, 102–103
Investigation, 80

J

Job posting, 101

K

Key personnel, 65–66, 70
Knowledge, skills, and abilities (KSA), 101, 105
Kyma, 126–129

L

Leadership models, 9–16
Leading indicators, 111, 116
Line management responsibility, 112–113
Lock-in mechanism, 85
Loyalty, 123

M

Make-buy decision, 56
Market environment, 2, 61, 63, 67
Marketing, 83–89, 130
Market research, 85, 126
Matrix organization, 120
Meetings, 75–77
Metrics, 41
Moore's Law, 34

N

Negotiations, 77–79, 129
Networks, 2, 3, 5–7

O

Operating risks, 111–116
Opportunity assessment, 62
Opportunity cost, 55
Opportunity environment, 6
Organizational change, 16–18
Organizational networks, 118–121
Organizational reward structure, 24, 25
Organogram (organization chart), 117–120
Overhead, 50, 101

P

Patent, 89, 125–126
Pay equity, 107
Perceptions of management, 91–92
Performance evaluation, 103–106
Planning retreat, 38–42
Positioning (organizational), 43–44, 60, 83, 88, 101, 103, 107
Positioning (products), 86–87
Price point, 84, 87
Product line, 83–85
Program reviews, 111–112
Project, 119

R

Ranking employees, 107
Research and development (R&D)
    definitions of, 19, 20
    funding levels, 21
    life cycle, 22
    management space, 23
    strategy, 27–28
Resource-loaded schedule, 46–47
Resource management, 46–48, 56–58
Responsibility, 117, 121–122
Rights of the manager, 18
Risk management, 19, 29, 30, 42, 111–116
Risk registry, 31, 35
Root-cause analysis, 112, 116

S

Salary incentives, 108–109
Salary management, 106–109
Science policy, 24
Scientific misconduct, 93–95
S-curves, 22
Sincrotrone Trieste, 127
Skills matrix, 100
Spin-off, 126–130
Staffing mix, 99–100
Staff retention, 106
Stop work authority, 115
Storyboard, 45, 67, 74
Strategic marketing, 88–89
Strategic partnerships, 89, 129–130
Strategic plan, 37–48, 123, 129
Stretch assignments, 105
Succession planning, 103–106, 123

T

Technical writing, 69–81
Technology transfer, 125–130

V

Value chain, 61
Value engineering, 55
Value proposition, 61, 63
Value space (of the enterprise), 24

Virtual company, 127–129
Vision and mission statements, 1, 4, 23,
    39, 43

W

Writing the plan, 44–46
Work breakdown structure, 34, 46–47,
    72, 100
Workforce management, 99–109
Work package, 54